I0064144

DESIGN OF PILE FOUNDATIONS IN LIQUEFIABLE SOILS

DESIGN OF PILE FOUNDATIONS IN LIQUEFIABLE SOILS

Gopal Madabhushi
University of Cambridge, UK

Jonathan Knappett
University of Dundee, UK

Stuart Haigh
University of Cambridge, UK

Imperial College Press

Published by

Imperial College Press
57 Shelton Street
Covent Garden
London WC2H 9HE

Distributed by

World Scientific Publishing Co. Pte. Ltd.
5 Toh Tuck Link, Singapore 596224
USA office: 27 Warren Street, Suite 401-402, Hackensack, NJ 07601
UK office: 57 Shelton Street, Covent Garden, London WC2H 9HE

British Library Cataloguing-in-Publication Data
A catalogue record for this book is available from the British Library.

DESIGN OF PILE FOUNDATIONS IN LIQUEFIABLE SOILS
Copyright © 2010 by Imperial College Press

All rights reserved. This book, or parts thereof, may not be reproduced in any form or by any means, electronic or mechanical, including photocopying, recording or any information storage and retrieval system now known or to be invented, without written permission from the Publisher.

For photocopying of material in this volume, please pay a copying fee through the Copyright Clearance Center, Inc., 222 Rosewood Drive, Danvers, MA 01923, USA. In this case permission to photocopy is not required from the publisher.

ISBN-13 978-1-84816-362-1
ISBN-10 1-84816-362-2

Printed in Singapore.

From Gopal Madabhushi:

To my parents for allowing me to pursue what I wanted and to Raji, Spandana and Srikanth for reminding me of the *meaning of life*.

From Jonathan Knappett:

To Lis, for the unwavering support you continue to give me and to my parents, for making me the person I am today.

From Stuart Haigh:

To Esra and Maya for their constant support and the joy they bring to my life.

Foreword

The Design of Pile Foundations in Liquefiable Soils provides a systematic evaluation of the effects of liquefaction on pile foundations and includes earthquake observations, research findings, and design principles and procedures from a variety of sources worldwide. The book provides a logical framework for understanding the basics of single pile and pile group design, liquefaction, and the effects of earthquake loading and liquefaction on the axial and lateral loads transmitted to pile foundations. It also provides a framework for understanding the effects that loss of bearing and lateral restraint in saturated sandy soils subject to cyclic loading have on the capacity of pile foundations. By combining earthquake loading in liquefiable soils with mechanisms that reduce pile capacity, the book develops a rational process for quantifying loads and capacity reduction into a design process.

Pile response to earthquakes and liquefaction involves complex material behavior in terms of increased pore pressure and reduced soil strength and stiffness, complex mass behavior of the ground in terms of kinematic loading, complex performance of the superstructure in terms of inertial loads transmitted to the piles, and complex interaction between the soil and pile foundation. This book provides a valuable guide for students, researchers, and designers in navigating these complexities.

Chapter 1 focuses on the performance of pile foundations with a review of factors contributing to axial pile capacity, performance-based design principles for piles subject to earthquakes, and the observed performance of pile foundations during previous earthquakes. Chapter 2 reviews inertial and kinematic loading, and presents the essentials of p-y analysis and limit equilibrium methods to estimate the maximum shear and moment in piles in laterally spreading soils. Chapter 3 covers factors affecting the axial loading of piles in liquefiable soils, including reduction in end-bearing capacity due to elevated pore pressures and the potential for beam buckling associated with the loss of lateral restraint in liquefying soils. Chapter 4 focuses on lateral spreading in liquefied soils, including empirical methods for estimating lateral spread movement, soil-pile interaction in liquefied soil, and limiting lateral earth pressures for pile design. Chapter 5 evaluates combined axial and lateral pile loading effects in laterally spreading ground with a treatment of single pile and pile group behaviour and the use of interaction diagrams to analyse pile performance when multiple failure mechanisms are possible. Chapter 6 provides a substantial number of design examples to illustrate the design procedures developed in the book.

The Design of Pile Foundations in Liquefiable Soils also demonstrates the importance of centrifuge testing to identify and quantify key failure mechanisms associated with complex soil-structure interaction. The combined use of careful field observations, centrifuge experiments and fundamental mechanics to develop analytical procedures and a structured design process is well illustrated by this book. The provision of design examples is a particularly attractive feature. The worked examples show explicitly how to apply the design process, and provide an indispensible link between theory and practice.

The Design of Pile Foundations in Liquefiable Soils provides substantial forward progress in conceptualising and formalising the analytical and design treatment of the truly complex, and sometimes vexing, phenomena associated with pile response to earthquake loading and liquefaction. For those interested in liquefaction, liquefaction-induced soil-structure interaction, pile design, and centrifuge testing, this book will be a valuable and frequently used addition to their technical library.

T.D. O'Rourke
Thomas R. Briggs Professor of Engineering
School of Civil and Environmental Engineering
Cornell University
Ithaca, NY USA

12 April 2009

Preface

Pile foundations are widely used both onshore and offshore to transfer superstructure loads into the ground. In seismic regions there is uncertainty regarding their performance, particularly when the soil strata that the piles pass through or bear on are susceptible to liquefaction. This book aims to clarify the mechanisms by which pile foundations may fail when the soil suffers liquefaction. In addition, the problem of down slope movement of nonliquefied ground that overlies liquefied layers and its effect on pile foundations is considered. One of the key factors to remember is that the pile foundations are often carrying substantial axial loads from the superstructure at the time of the earthquake. It is therefore imperative to consider the pile behaviour in liquefiable soils when they are subjected to both axial and lateral loads. This can have consequences for pile behaviour such as sudden and catastrophic buckling instability or excessive and unwarranted settlements.

This book is organised into six chapters. It aims to marry the most recent research findings on pile behaviour to the needs of practical designs in seismic regions. Accordingly, it is intended to be used by graduate students and researchers interested in pile foundation design, as well as geotechnical practitioners faced with the problem of designing or assessing the seismic risk to existing pile foundations in regions with liquefiable soils.

Chapter 1 introduces the static design of piles using traditional or CPT-based methods. It looks at the performance of piles in past earthquakes through some well-documented case studies. The concepts of performance design and the importance of estimating deformations of pile foundations are highlighted.

Chapter 2 deals with the inertial and kinematic loads attracted by pile foundations during earthquake loading. These are considered initially for normal ground and later on in liquefiable soils. Limit equilibrium-based methods are introduced to estimate the loading due to laterally spreading nonliquefied layers.

Chapter 3 introduces liquefaction as a foundation hazard and discusses how the consequent loss of soil strength influences the axial load that can be safely carried by pile foundations. Possible axial failure modes including liquefaction-induced bearing capacity failure and instability (buckling) are discussed in relation to the static considerations outlined in Chapter 1.

Chapter 4 discusses the lateral spreading of sloping ground and the particular problems that arise when pile foundations pass through such laterally spreading ground. Recent research experiences are presented and compared with current codal provisions.

Chapter 5 brings together the material in Chapters 2 to 4 in considering the design of pile foundations against combinations of transient seismic loads (Chapter 2), axial load (Chapter 3) and kinematic forces due to lateral spreading (Chapter 4).

Chapter 6 presents a series of design examples to demonstrate how the methods outlined in Chapters 1 to 5 may be combined within an inclusive design method.

Finally, the authors would like to thank Professor Tom D. O'Rourke of Cornell University, USA and the 49th Rankine Lecturer, for writing the foreword. Similarly the authors would like to acknowledge the help and support of many researchers at the Schofield Centre who shared in the enthusiasm of understanding the complex problems of soil liquefaction and earthquake geotechnical engineering. Particular mention must be made of the excellent support received from the technical staff during many an experimental project. In fact, it is the excellent research atmosphere at the Centre of freely sharing knowledge and technical know-how that makes work a pleasure and made this book possible.

Contents

Chapter 1

Performance of Pile Foundations

1.1 Introduction

Pile foundations are the most popular form of deep foundations used for both onshore and offshore structures. They are often used to transfer large loads from the superstructure into deeper, competent soil layers particularly when the structure is to be located on shallow, weak soil layers. There are many ways by which pile foundations can be classified. For example, by material (steel or reinforced concrete piles) or by method of installation (driven, jacked or bored piles). Pile foundations can be classified based on their functionality and the mechanism by which they carry the load as shown schematically in Fig. 1.1 and described below:

- *End bearing piles*: the superstructure load is transferred through water or weaker soils onto a competent bearing stratum by means of pile tip resistance (Fig. 1.1a).
- *Friction piles*: the superstructure load is transferred into the soil through the frictional resistance generated along the shaft of the pile. Sometimes this type of piles are called *floating piles* as their end bearing may be neglected (Fig. 1.1b).
- *Compaction piles*: piles driven into loose, granular soils to make the soil more dense by displacing it. Normally such piles are not used for load carrying purposes and may simply involve driving a steel tube into the ground that is withdrawn while replacing the tubular volume with granular material to form a 'sand pile' (Fig. 1.1c).
- *Tension piles*: piles that are employed to resist pull-out forces either in the vertical or inclined directions. Such piles are often

used where the superstructure is expected to experience lateral forces either due to earthquake or wind action (Fig. 1.1d) or in the case of offshore foundations where the piles are used to anchor tethers, cables etc (Fig. 1.1e).

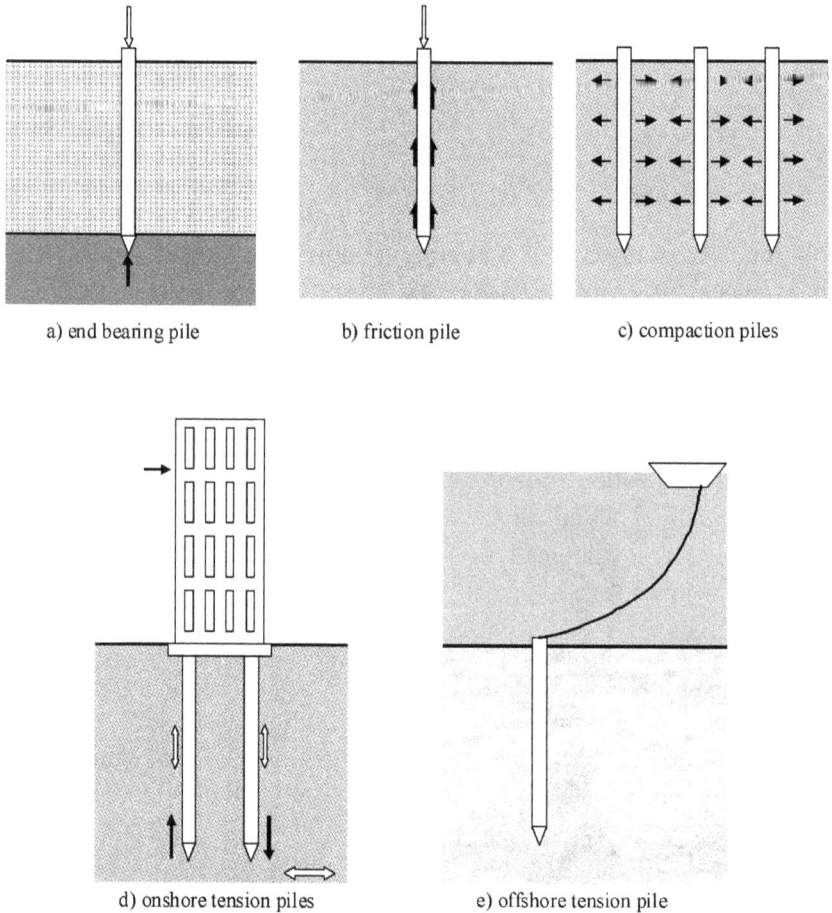

a) end bearing pile   b) friction pile   c) compaction piles

d) onshore tension piles   e) offshore tension pile

Fig. 1.1 Classification of pile foundations.

In this book the primary emphasis is on the design of pile foundations subjected to earthquake loading. Therefore, the focus will be on the pile foundations in saturated, granular soils that are vulnerable due to liquefaction. There are many case histories that show pile foundations that suffer severe damage or fail completely when the ground surrounding the pile liquefies. This aspect will be explained in sections 1.3 and 1.5 by considering in detail some case histories on poor performance of pile foundations following liquefaction.

1.1.1 Axial capacity of a single pile

For a single pile, the axial load carrying capacity can be written from vertical equilibrium considerations as

$$Q_u = Q_b + Q_s \qquad (1.1)$$

where Q_u – axial capacity of the pile, Q_b – base resistance at the pile tip and Q_s – shaft friction of the pile. In general terms, the base resistance Q_b can be calculated as

$$Q_b = A_b \sigma_b \left(N_q - 1 \right) \qquad (1.2)$$

where A_b – base area of the pile, σ_b – effective overburden pressure at the pile tip level and N_q – bearing capacity factor that can be estimated using Fig. 1.2 below following Berezantsev *et al.* (1961), based on the angle of shearing resistance of the soil.

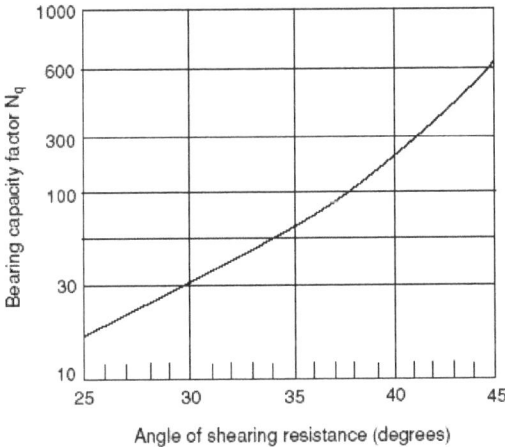

Fig. 1.2 Bearing capacity factor N_q for deep foundations. (Berezantsev *et al.*, 1961).

The shaft capacity may be obtained by estimating the shear stress generated along the shaft which can be calculated as

$$\tau_s = K_s\, \sigma'_v\, tan\delta_{cv} \tag{1.3}$$

where K_s – is a earth pressure coefficient, σ'_v – is effective vertical stress at a given elevation and δ_{cv} – friction angle between the pile material and the soil. K_s depends on type of pile and installation method (driven or cast *in situ* piles). Broms (1966) related the values of K_s and δ_{cv} to the angle of shearing resistance of the soil ϕ' as shown in Table 1.1.

In order to obtain the shaft capacity due to skin friction, the shear stress must be integrated over the surface area of the pile using the following equation,

$$Q_s = 2\pi r \times \int_0^L \tau_s \tag{1.4}$$

where r is the pile radius and L is the length of the pile.

Table 1.1 Values of earth pressure coefficient K_s and pile-soil friction angle δ. (Broms, 1966).

Pile Material	δ_{cv}	K_s	
		Low relative density	High relative density
Steel	20°	0.5	1.0
Concrete	0.75 ϕ'	1.0	2.0
Wood	0.66 ϕ'	1.5	4.0

1.1.2 Pile capacity based on CPT testing

It is a common practice to estimate the axial capacity of a pile based on the cone resistance obtained from the standard cone penetration tests (CPTs) carried out as part of the site investigation. In his Rankine Lecture, Randolph summarised the design of pile foundations in cohesive and granular soils based on state-of-the-art methods. (Randolph, 2003). With regard to piles in granular soils, Eqs. 1.2 and 1.4 suggest an increase in the axial capacity of the pile with increasing depth of penetration. However, it is well recognised in the literature that both the base resistance and shaft friction reach limiting values at some 'critical depth' of penetration. The critical depth may be expressed in absolute terms or be normalised by the pile diameter. (Vesic 1967, 1970; Coyle and Castello, 1981).

The base resistance may reach a limiting value for two reasons. Firstly the angle of shearing resistance may decrease with increasing stress level, i.e. with increasing depth of pile penetration. For example, following Bolton (1986), the angle of shearing resistance will depend on the critical state friction angle and the angle of dilatancy. Although the critical state friction angle remains constant, with increasing stress level the angle of dilatancy can become smaller thereby leading to a reduced angle of shearing resistance. Cheng (2004) modified the Berezantzev *et al.* method to account for the angle of dilatancy determined following the work of Bolton (1986).

The second reason for a reduction in base resistance at greater depths of pile penetration is the nonlinear relationship between stiffness and stress. Both these factors suggest that the base resistance may reach a limiting value. Similarly the shaft friction is thought to reach a limiting value as the normal effective stress is found to degrade owing to installation of the pile that causes a gradual densification of the granular material in the vicinity of the pile. More recent research (Zhao, 2008) on installation of miniature CPTs in centrifuge models indicates that the increase in shaft friction may be more due to the increased horizontal stresses rather than the changes in soil density in the vicinity of the pile.

1.1.2.1 Pile base capacity

It is logical to imagine that the mechanisms at work during the installation of a pile are similar to those at work while driving a CPT into the ground. However, due consideration must be given to the displacements required to mobilise a proportion of the cone resistance during a CPT test verses the displacements required to mobilise a portion of the base resistance of a pile. For bored piles, Fleming (1992) suggests the following hyperbolic relationship between end bearing pressure q_b and the base displacement w_b

$$\frac{q_b}{q_c} \approx \frac{w_b/d}{w_b/d + 0.5\, q_c/E_b} \tag{1.5}$$

where d is the pile diameter, q_c is the cone resistance and E_b is the Young's modulus of the soil below the pile base.

For driven piles, a degree of 'locked in' or residual pressure q_{b0} is to be expected at the base of the pile. Eq. 1.5 can be modified to include this effect as follows:

$$\frac{q_b}{q_c} \approx \frac{w_b/d + 0.5\, q_{b0}/E_b}{w_b/d + 0.5\, q_c/E_b} \tag{1.6}$$

The ratio of residual pressure to cone resistance q_{b0}/q_c can be conservatively taken to have a range of 0.3–0.7.

In recent work at Imperial College, Chow (1997) analysed a database of high-quality pile load tests and CPT tests. Based on this analysis, Jardine and Chow (1996) suggest the following relationship between the end bearing pressure q_b and the cone penetration resistance q_c,

$$\frac{q_b}{q_c} = 1 - 0.5 log\left(\frac{d}{d_{cone}}\right) \geq 0.13 \tag{1.7}$$

where d is the diameter of the pile and d_{cone} is the cone diameter.

For a nominal cone diameter of 25.4 mm, Eq. 1.7 above can be used to plot q_b/q_c verses pile diameter as shown in Fig. 1.3.

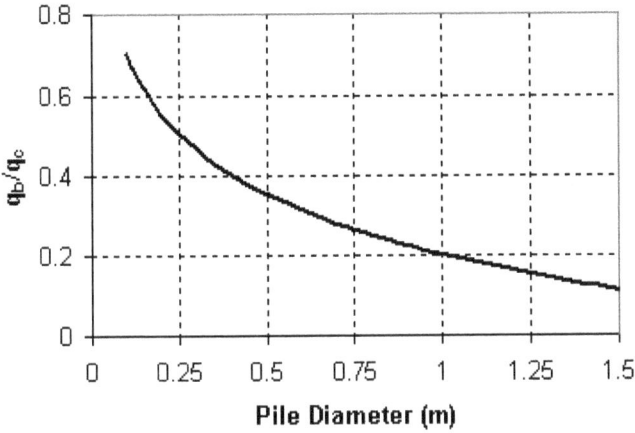

Fig. 1.3 Variation of normalised end bearing with pile diameter. (Jardine and Chow, 1996).

1.1.2.2 Shaft friction

The shaft friction of a pile driven into granular soils can be estimated at any depth in terms of the cone resistance q_c using Eq. 1.8 below, which was derived following the MTD method developed at Imperial College based on high-quality pile tests. (Jardine and Chow, 1996).

$$\tau_s = \left[\frac{q_c}{45} \left(\frac{\sigma'_{v0}}{p_a} \right)^{0.13} \left(\frac{d}{h} \right)^{0.38} + \Delta\sigma'_{rd} \right] tan\delta_{cv} \qquad (1.8)$$

where p_a is atmospheric pressure, σ'_{v0} is the effective stress at the given depth, d is the pile diameter, h is the height to the soil surface, $\Delta\sigma'_{rd}$ is the stress increase due to dilation of the sand from passage of pile tip at the current depth to the current position of the pile and δ_{cv} is the interface friction angle.

An alternative method for estimating the shaft friction at any depth was suggested by Randolph *et al.* (1994). This method assumes that the radial stresses around the pile decrease exponentially with radial distance away from the surface of the pile shaft. The shear stress on the pile shaft can be estimated using

$$\tau_s = \left[K_{min} + \left(K_{max} - K_{min} \right) e^{-\mu h/d} \right] \sigma'_{v0} \, tan\, \delta_{cv} \qquad (1.9)$$

where σ'_{v0} is the effective stress at the given depth, K_{max} is taken proportional to the cone resistance q_c, normalised with the effective stress, normally 1–2% of (q_c/σ'_{v0}), K_{min} is taken in the range of 0.2–0.4 and μ is taken as 0.05 for typical pile diameters. For large pile diameters this value may have to be reduced suitably.

This method gives comparable results to the MTD method of Jardine and Chow (1996) outlined in Eq. 1.8. The main difference is that the MTD method assumes a power law for the decrease in radial stress with increasing distance from the pile shaft while the method outlined in Eq. 1.9 assumes an exponential decrease.

It must be pointed out that while considering the cone resistance q_c at a stratified site with soil layers of contrasting stiffness the cone resistance may have to be weighted to reflect the weaker materials, for example following Meyerhof and Valsangkar (1977).

1.2 Performance of Pile Foundations During Earthquake Loading

In the above section the general methodology for estimating the axial pile capacity is presented. The basic assumption employed is that the load applied to the pile at its head is carried by the sum of shaft resistance and end bearing. Both end bearing and shaft resistance increase nonlinearly with depth, i.e. the increased axial capacity of the pile below a critical depth of penetration is rather limited. The end bearing and shaft resistance may be estimated based on CPT tests conducted at the site as outlined in section 1.1.2 above.

If a single pile is loaded to failure, for example during a pile load test, then the limiting values of the end bearing and shaft resistance are fully mobilised. However, when a pile is subjected to working loads the proportion of end bearing compared to shaft resistance that is mobilised may vary. This may also be a function of the method of construction employed, for example driven or jacked-in pile versus cast *in situ* or bored pile.

The actual end bearing and shaft resistances mobilised will depend on the displacements suffered by the pile under the working loads applied. This suggests that geotechnical engineers need to have a good idea of the displacements mobilised by a pile under working loads. In addition to this the modern trend to move towards performance-based design requires the geotechnical engineer to be able to estimate displacements suffered by the piles under various load combinations so that the performance criterion can be set.

Dynamic loading due to earthquakes will be superimposed onto the working loads already acting on the pile. Therefore, when designing piles in an earthquake-prone area a clear distinction needs to be made regarding the expected performance of the piles. It may be a design requirement that the pile should only suffer small displacements during a more probable but moderate earthquake, i.e. the pile needs to perform to meet serviceability requirements. Following the seismic design guidelines by PIANC (2001) and the more recent ISO 23469 guidelines, such an earthquake motion may be called a Level 1 (L1) earthquake, which will have a probability of displacement of 50% during the design life of the structure. On the other hand, during a less probable but stronger earthquake the pile may be allowed to suffer more displacement but must perform adequately to prevent total collapse of the superstructure. Such an earthquake motion would be classed as Level 2 (L2) earthquake motion which will have a probability of displacement of 10% during the design life of the structure. On average an L1 earthquake has a return period of about 72 years while a L2 earthquake has a return period of about 475 years.

The performance guidelines for any geotechnical structure may be set based on an acceptable level of damage following the PIANC guidelines. The PIANC guidelines were developed with the design philosophy of protection of human life and property with due consideration of the structure's ability to provide emergency services during the post-earthquake period and prevention of any spillage of hazardous materials (this may be more appropriate to port and harbour structures). In Table 1.2 the acceptable level of damage is described in relation to serviceability (SLS) and ultimate limit states (ULS).

Table 1.2 Acceptable level of damage in performance-based design. (PIANC, 2001).

Level of damage	Structural	Operational
Degree I: Serviceable	Minor or no damage	Little or no loss of serviceability
Degree II: Repairable	Controlled damage	Short-term loss of serviceability
Degree III: Near collapse	Extensive damage in near collapse	Long-term or complete loss of serviceability
Degree IV: Collapse	Complete loss of structure	Complete loss of serviceability

If performance-based design philosophy is followed then piled foundations are to be designed by selecting a performance grade. PIANC guidelines provide for four grades of performance: S through C as explained in Table 1.3. The performance grade for a pile foundation may be chosen based on the importance of the superstructure as defined by local seismic codes and standards and the requirements of the users/ operators of the superstructure. Schematically the performance of different grades during L1 and L2 earthquake events be classified as shown in Fig. 1.4.

Table 1.3 Performance grades S, A, B and C.

| Performance grade | Design earthquake | |
	Level 1 (L1)	Level 2 (L2)
Grade S	Degree I: serviceable	Degree I: serviceable
Grade A	Degree I: serviceable	Degree II: repairable
Grade B	Degree I: serviceable	Degree III: Near collapse
Grade C	Degree II: repairable	Degree IV: Collapse

Based on the performance grade chosen for the pile foundation, the limits of displacement and/or deformation may have to be set to keep the level of damage to the required levels (serviceable to collapse) during L1 and L2 earthquake events. These may be different for different types of deformation. For example, the limits set for vertical settlement of the pile, horizontal displacement at pile cap level and the allowable rotation at the pile head may be chosen considering the performance grade of the superstructure during the L1 and L2 events. The geotechnical engineer is then faced with the challenge of designing the pile foundation to limit these displacements and deformations of the pile to achieve the required performance. This concept is quite different from the traditional load-/stress-based factor of safety approach.

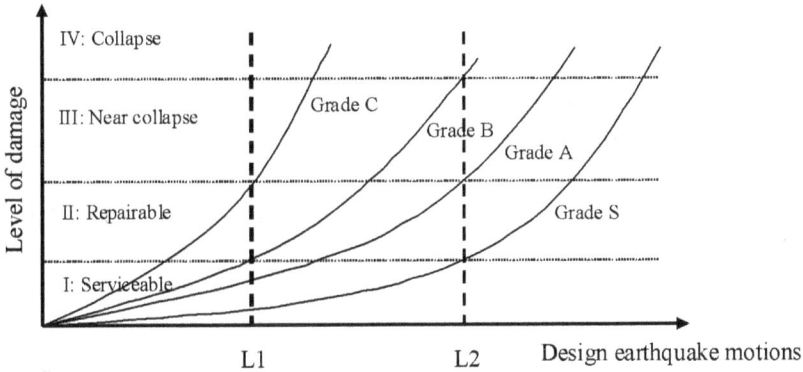

Fig. 1.4 Schematic diagram of performance.

1.3 Soil Liquefaction and Lateral Spreading

When loosely deposited sandy soil layers are subject to shaking from earthquake loading, the soil grains tend to reorganise into a denser packing thereby exhibiting a volumetric contraction. If the soil layers are fully saturated then the contractive tendency of the soil layers is manifested as a rise in the pore water pressure as there is insufficient time for the pore water to escape from the soil layers. If the excess pore water pressure increases sufficiently to match the effective stress in the soil, then it is said that the soil layers have suffered full liquefaction. If the excess pore water pressures that are generated in the soil due to the earthquake loading are limited, then the soil may still soften and the soil is said to have suffered partial liquefaction.

The foundations of civil engineering structures suffer settlement and/or rotation when the soil layers suffer liquefaction. Pile foundations can be particularly vulnerable to soil liquefaction. If the soil below the base of the pile becomes liquefied, then there is a decrease in the base capacity and the pile can suffer excessive settlements. On the other hand, if the depth to which soil liquefies is rather limited, say in a relatively small magnitude earthquake, the soil surrounding the shaft may liquefy and a loss of shaft friction may be expected. This can cause an increase in the base load of the pile, which can lead to an increased settlement. In addition to these simple cases, pile foundations are also vulnerable to laterally spreading ground.

Lateral spreading of sloping ground can occur if the soil layers suffer either full or partial liquefaction. When excess pore pressures are generated in a sloping layer, the slope may no longer be able to resist the static shear stress and can start to flow. Pile foundations that are passing through a laterally spreading ground will therefore be subjected to lateral kinematic loads. Generally, if the soil layer has fully liquefied then the lateral forces exerted on the piles by flowing ground may be considered to be relatively small. However, if there is a stiff soil layer above the liquefied layer, such a layer would also spread. These nonliquefied layers can exert large lateral loads on pile foundations as they are able to generate significant passive pressures against the piles and/or pile cap.

1.4 Performance of Pile Foundations in Past Earthquakes

Historically there have been many case histories where pile foundations have suffered either total collapse or severe damage during earthquake loading. In Table 1.4 details of case histories of pile foundation behaviour from sixteen different sites are assembled in various strong earthquake events. Some of the pile foundations performed adequately while others have suffered severe damage as indicated in Table 1.4.

In this section four of the prominent case histories are discussed briefly. In the next section the main lessons that have been learnt from these failures are presented.

Performance of Pile Foundations   13

Table 1.4 Summary of case histories on pile foundation performance in past earthquakes.

No.	Case history	Earthquake event	Pile material	Pile diameter (m)	Pile length (m)	Lateral spreading observed?	Pile performance	Reference
1	10 storey-Hokuriku building	Niigata earthquake 1964	RCC	0.4	12	Yes	Good	Hamada (1992a,b)
2	Showa bridge	Niigata earthquake 1964	steel tubular	0.6	25	Yes	Poor	Hamada (1992a,b)
3	Landing bridge	Edgecumbe earthquake, 1987	PSC (square)	0.4	9	Yes	Good	Berrill et al. (2001)
4	14 storey building in American park	Kobe earthquake, 1995	RCC	2.5	33	Yes	Good	Tokimatsu et al. (1996)
5	Hanshin expressway pier	Kobe earthquake, 1995	RCC	1.5	41	Yes	Good	Ishihara (1997)
6	LPG tank 101	Kobe earthquake, 1995	RCC	1.1	27	Yes	Good	Ishihara (1997)
7	Kobe Shimim hospital	Kobe earthquake, 1995	steel tubular	0.66	30	Yes	Good	Soga (1997)
8	NHK building	Niigata earthquake 1964	RCC	0.35	12	Yes	Poor	Hamada (1992a)
9	NFCH building	Niigata earthquake 1964	RCC hollow	0.35	9	Yes	Poor	Hamada (1992a)
10	Yachiyo Bridge	Niigata earthquake 1964	RCC	0.3	11	Yes	Poor	Hamada (1992a,b)
11	Gaiko Ware House	Chubu earthquake, 1983	PSC hollow	0.6	18	Yes	Poor	Hamada (1992b)
12	4 storey fire house	Kobe earthquake, 1995	PSC	0.4	30	Yes	Poor	Tokimatsu et al. (1996)
13	3 storied building, Fukae at Kobe	Kobe earthquake, 1995	PSC	0.4	20	Yes	Poor	Tokimatsu et al. (1998)

Table 1.4 (*Continued*)

14	Elevated port liner railway	Kobe earthquake, 1995	RCC	0.6	12	Yes	Poor	Soga (1997)
15	LPG tank – 106,107	Kobe earthquake, hollow 1995	RCC	0.3	20	No	Poor	Ishihara (1997)
16	Harbour Master's building, Kandla port	Bhuj earthquake, India, 2001	RCC	0.4	25	Yes	Poor	Madabhushi *et ul.* (2003)

RCC – Reinforced concrete
PSC – Prestressed concrete

1.4.1 Showa bridge failure

The Showa bridge collapse is an interesting case history and to date evokes passionate discussions among geotechnical engineers. This case history happened during the Niigata earthquake of 1964, which had a Richter magnitude of 7.5 or between VII and VIII on a Modified Mercalli scale. The earthquake had a focal depth of 40 km and the epicentre was near Awa-shima island, some 22 km off the coast in the Sea of Japan.

The Showa bridge spanned the Shinano river and had ten main spans of 28m each with two smaller spans onto the abutments. The bridge was only completed five months before the Niigata earthquake struck. The bridge decks were simply supported on movable bearings at one end and fixed at the other except for the middle pier where both the decks supported on the pier were on movable bearings. The bridge was slightly unusual in that the piles that were driven into the ground extended above the water to form the pier. Also all the piles at any pier location were positioned along a single line. As a result, the bridge had considerable flexibility in the longitudinal direction.

The Niigata earthquake caused widespread liquefaction in the city of Niigata and in the surrounding areas. The banks of the Shinano river suffered lateral spreading following liquefaction, particularly the left bank, which moved by about 5m. The Showa bridge suffered catastrophic failure as shown in Fig. 1.5. A schematic diagram showing the collapsed decks and bent piles is shown in Fig. 1.6. The reasons for the collapse of the bridge were initially considered to be due to the dynamic response of the bridge, especially as pier P6 had different support conditions (both decks resting on it were on movable bearings) compared to other piers.

Hamada (1992a,b) argued that a more plausible explanation could be offered based on the ground displacements suffered due to liquefaction-induced lateral spreading. In Fig. 1.7 the schematic view of a pile section extracted from location P4 is shown along with the soil profile and SPT values at the bridge site. As the left bank has suffered lateral spreading of about 5m towards the centre of the river channel, the piles are subjected to horizontal loading induced by the lateral earth pressures. The piles then deform in the longitudinal direction of the bridge, causing the decks to dislodge. The observation of the bent pile shown in Fig. 1.7 extracted after the event gives credence to this type of failure. In addition the eye-witness accounts described the falling of decks to have taken place some time after the earthquake shaking had stopped (between 0 and ~1 minute). Following the Niigata earthquake many researchers concentrated on this lateral loading induced by spreading of liquefied ground and extensive research was carried out to investigate the behaviour of piles under such loading conditions. The JRA code (1996, 2002) tried to formalise this research and presented methods of estimating the loading due to lateral spreading ground on pile foundations.

More recently the failure of Showa bridge was revisited by Bhattacharya *et al.* (2005). Based on the work of Bhattacharya *et al.* (2004) it was argued that the Showa bridge could have collapsed due to the buckling of pile foundations rather than due to lateral spreading of ground causing excessive lateral displacements of the piles at the bridge site. The buckling or instability mechanism of failure for piles in liquefied ground will be discussed in detail in Chapter 3. Following the work of Bhattacharya and his co-workers, renewed interest was created in the collapse of the Showa bridge. For example, Yoshida *et al.* (2007) have revisited this and collated a number of eye-witness accounts to establish the timing of the bridge collapse as well as the lateral spreading of the river banks. Based on this work they conclude that the possibility of the bridge collapsing due to inertial loading or liquefaction induced soil flow is rather low. They conclude that the Showa bridge collapse was most likely due to the increased displacement of the ground owing to liquefaction wherein the pile deformation can occur more easily. Motoki *et al.* (2008) also revisited this problem recently and used analytical methods to explain the collapse of the Showa bridge. They obtained good comparisons of the pile deformation (see Fig. 1.7) but assume that the loading on the pile is due to lateral spreading ground.

From a performance point of view, it can be said that the magnitude 7.5 earthquake was probably close to an L2 event for the Showa bridge. The bridge suffered total collapse and complete loss of serviceability following the earthquake, therefore its performance grade would be between B and C (referring to Fig. 1.4).

Fig. 1.5 A view of the collapsed Showa bridge (Photo by V V Bertero) Courtesy of National Information Service for Earthquake Engineering EERC, University of California, Berkeley.

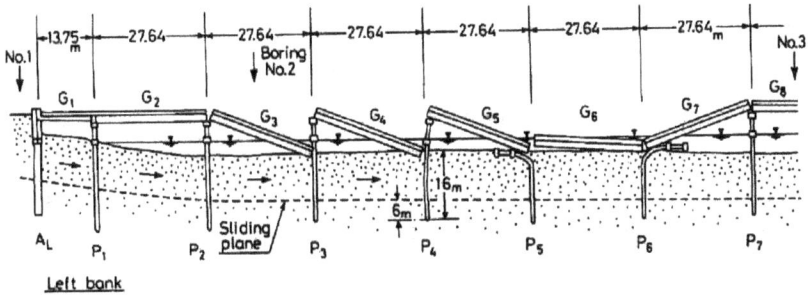

Fig. 1.6 Schematic diagram showing the fallen decks of the Showa bridge. (Takata *et al.*, 1965).

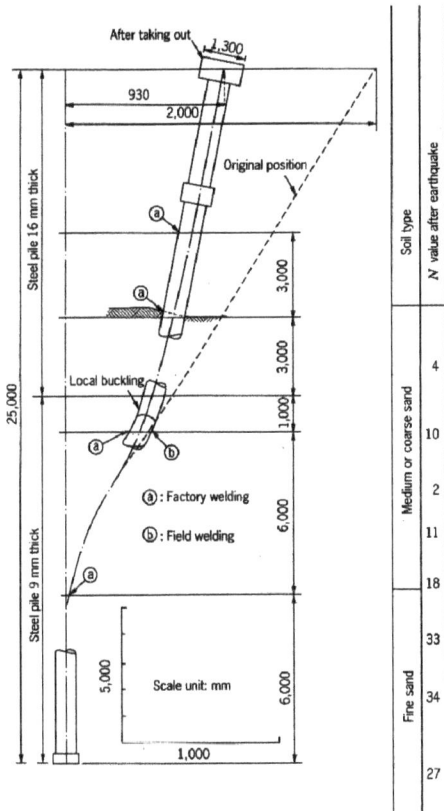

Fig. 1.7 Pile extracted from P4 location during post earthquake evaluation. (Fukuoka, 1966).

1.4.2 Niigata Family Court House building

The Niigata Family Court House (NFCH) building was a four-storey building supported on concrete pile foundations. The plan view of this 'L'-shaped building is presented in Fig. 1.8, which shows the footing locations and the connecting foundation beams. The NFCH building was constructed about one year before the Niigata earthquake of 1964. The building was located in the Hakusan area on the left bank of the Shinano river where vast lateral spreading was observed, for example, at the Showa bridge site described in Sec. 1.5.1 above. The concrete piles were 0.35m in diameter and between 6m and 9m long as indicated in Table 1.4. Hamada (1992a,b) reports horizontal ground displacements of about 1.5m close to the NFCH building while the building itself suffered horizontal displacement of about 1m.

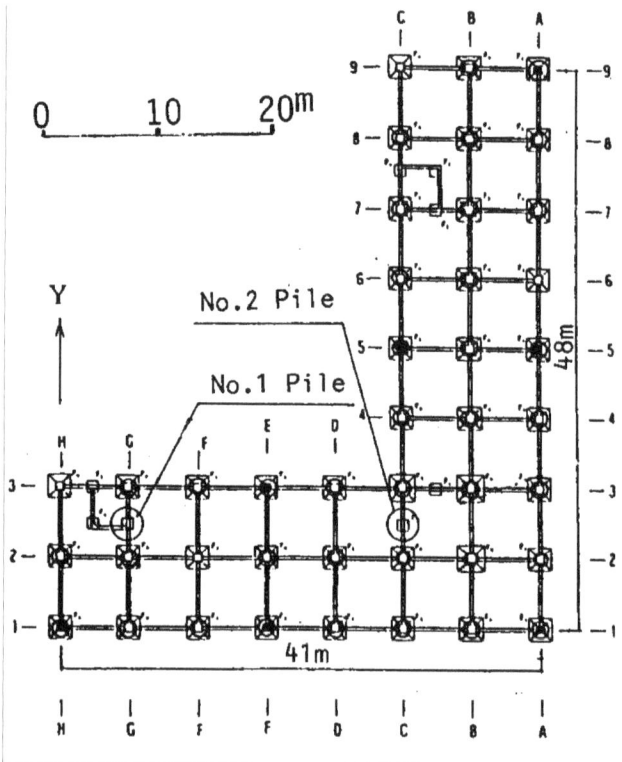

Fig. 1.8 Plan view of NFCH building showing the locations of extracted piles.

Following the Niigata earthquake the piles below the NFCH building suffered differential settlements. As a result the building was inclined by about 1° to the vertical. Damage to the pile foundations was suspected. However, minor repairs were carried out on the inclined floors and after that the building was used for another 25 years. The building was subsequently reconstructed. During the reconstruction two of the original piles were extracted, one of 6m length (Pile 1) and one of 9m length (Pile 2). The locations of these two piles in the original building are marked in Fig. 1.8. The damage to these piles was carefully recorded as reported by Hamada (1992a,b). Fig. 1.9 presents a schematic diagram showing the damage recorded in Piles 1 and 2. In the same figure the variation of SPT values with depth at this building site are shown. The shaded area in this figure below the water table indicates the estimated depth of liquefaction based on low SPT values of N <15. In Fig. 1.10 the photographs of the extracted Piles 1 and 2 are presented at the locations of distress marked on Fig. 1.9. From these photographs it can be seen that Pile 1 showed cracking in the top region only while Pile 2 showed severe cracking at the top and some cracking at the base. The major difference between these two piles is the depth of penetration of the piles into the ground. Pile 1 had a depth of penetration of about 6m while Pile 2 had a depth of penetration of 9m. Hamada (1992a,b) suggests that Pile 1 suffered relatively modest damage as it did not penetrate into the deeper, nonliquefied ground. As a result, the top of this pile could move with the building, while the base of the pile enjoyed relatively 'unconstrained' condition. In contrast, Pile 2 entered the deeper nonliquefied ground and therefore the base of this pile is relatively more constrained. As a result, when the NFCH building suffered lateral movement, the fixities at the top and bottom of Pile 2 caused bending at both of these locations, thereby exceeding the moment capacity of the pile at the top and at the base. This is supported by the damage presented in Fig. 1.10b and c. The main conclusion was that the laterally spreading ground around the piles caused the observed distress in these piles.

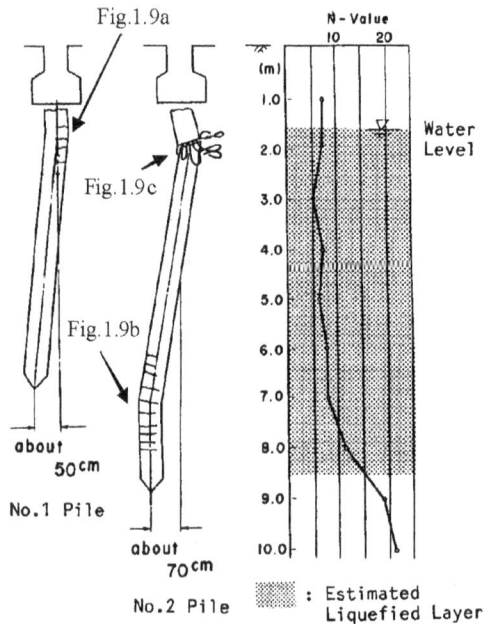

Fig. 1.9 Schematic diagram showing the distressed piles 1 and 2 after excavation.

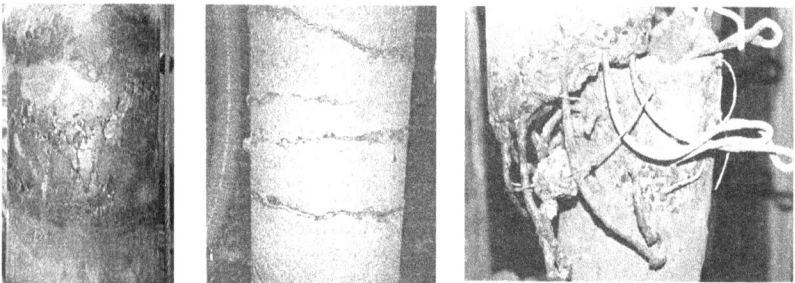

a) upper part of Pile 1   b) lower part of Pile 2   c) upper part of Pile 2

Fig. 1.10 Photographs of extracted piles showing damaged sections of Piles 1 and 2.

From a performance point of view, it is reasonable to suggest that the foundations of the NFCH building performed between Grade A and B (referring to Fig. 1.4) during an L2 earthquake event. The building was repaired following the earthquake and was subsequently used for a long period of time.

This case history also highlights some interesting questions. If the soil at the tip of Pile 1 was liquefied or lost some of its stiffness owing to excess pore water pressure generation, then could this pile have settled vertically, causing the NFCH building to rotate by 1°? Did Pile 2 suffer instability on the onset of liquefaction while being under the action of the superstructure load and lateral load due to the spreading ground? If the plastic moment capacity is exceeded at the top and bottom of Pile 2 as indicated in Fig. 1.8, could the NFCH building survive rotating by 1°? In other words, what made the building reach an equilibrium position at that point?

1.4.3 The Landing Bridge performance

The performance of the Landing Bridge in New Zealand during the 1987 Edgecumbe earthquake was discussed by Berrill *et al.* (2001). This earthquake had a magnitude of 6.3 on the Richter scale due to a rupture in the normal fault between the Pacific and Australian plates. The earthquake had its epicentre some 17km from the Landing road but the surface rupture was 8km from the Landing Bridge site. The peak ground acceleration of about 0.33g was measured at the Matahina dam site, which was also 8km from the surface rupture and hence this can be taken as the level of shaking at the bridge site.

The Landing Bridge runs across the Whakatane river and is on the road between Edgecumbe and Whakatane. The left bank of the river is the floodplain. An aerial view of the bridge is shown in Fig. 1.11. It was constructed in 1962 and had 13 spans, each 18.3m long. Each of the spans had five pre-cast 'I'-beams bearing on rubber pads of 16mm thickness. The beams were bolted to the piers forming stiff, moment resisting connections. The bridge piers were formed of concrete slabs running the full width of the bridge and supported on eight pre-stressed concrete piles 9m long and raked at 1H:6V. The piles each had square cross-sections of 0.406m. The bridge abutments were also supported by eight piles of the same cross-section in two rows, five piles on the riverside and three piles towards the approach. In 1984, the piers in the river channel were strengthened against scouring by adding two vertical, 1.1m diameter piles on as an extension to the existing pile cap. The soil profile and results of cone penetration tests at the site is presented in Fig. 1.12. The water table at the site is very shallow due to the proximity of the river. Therefore, the shallow layers of silty sands up to a depth of about 6m are all susceptible to liquefaction.

Fig. 1.11 An aerial view of the Landing Bridge.

The Edgecumbe earthquake resulted in vast lateral spreading at the bridge site and in its vicinity. Shallow soil layers moved towards the river channel from both sides, i.e. the left and right abutments, and this was confirmed by the observation of tension cracks at ground level. The pile foundations had to resist the additional horizontal loads induced by the laterally spreading soil layers. In Fig. 1.13 a view of Pier C is shown. This is a pier that is towards the left-hand abutment of the bridge in Fig. 1.11, located on the flood plain but outside the main river channel. In Fig. 1.13 it can be seen that the laterally spreading soil causes a passive bulge or heave against the pier on the right-hand side of the pier as it pushes against it and suffers a passive failure. On the left-hand side, the laterally spreading soil forms a gap, as it moves away from the pier.

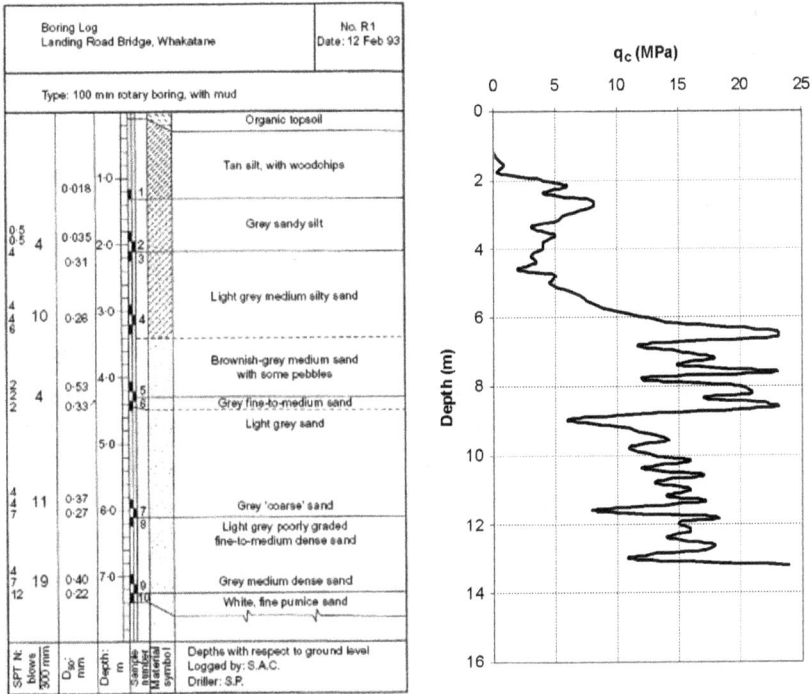

Fig. 1.12 A typical bore log at the Landing Bridge site.

Fig. 1.13 A view of Pier C in laterally spreading soil.

Berrill *et al.* (2001) report an extensive soil investigation carried out at the bridge site after the Edgecumbe earthquake. In Fig. 1.12 the cone penetration resistance q_c is re-plotted from the data presented by Berrill *et al.* for a location very close to Pier C shown in Fig. 1.13. Using a simplistic consideration that soil layers that have cone penetration resistance q_c <10 MPa are susceptible to liquefaction, it can be seen in Fig. 1.12 that the top 6m may have liquefied during the Edgecumbe earthquake, which was estimated to have had a PGA of 0.33g at the bridge site, for the reasons discussed above. It is important to note that the 9m-long raked piles would have the pile tips located at a depth of 8.8m where the q_c value was greater than 20 MPa.

Despite the lateral spreading and obvious liquefaction at the bridge site, the landing bridge survived the Edgecumbe earthquake with minor damage. The piles showed some signs of cracking at the top, just below the base of the pier. Berrill *et al.* present a detailed calculation of the plastic capacity of the piles and the bridge pier. They suggested a failure mechanism for the pile-pier system as shown in Fig. 1.14, in which plastic hinges form in the pile at the interface between the nonliquefied and liquefied soil layers, and at the top of the pile just below the pile cap. Similarly, the pier will have plastic hinges formed at the base of the pier and just below the bridge deck, assuming an integral connection between the pier and the deck due to the bolting method of connection employed in this bridge. The plastic moment capacities of the piles were estimated by Berrill *et al.* (2001) based on the structural drawings and material properties as 233kNm and 206kNm for the lower and upper hinge locations respectively. For the pier, the plastic moment capacities were 757kNm and 608kNm for the lower and upper hinge locations. These give a horizontal collapse load of about 1207kN. The actual horizontal load due to lateral spreading must have been somewhat smaller than this as the structure did survive the earthquake with minor damage. Berrill *et al.* (2001) report *in situ* shear box tests they conducted yielded a value for interlocking (c) as 10kPa and friction angle ϕ as 44° and unit weight of 12.5kN/m^3. These values yield with a traditional Coulomb analysis (with a wall friction of about 15°) a passive force of 2000kN, while a Rankine analysis yields a value of about 850kN. The authors comment on the possibility that the measured friction angle for the inherently loose granular material may be too large due to the presence of partially saturated silt layers. It is also possible that the Edgecumbe earthquake did not liquefy the soil to the assumed depth of 6m, which would have reduced the distance between the plastic hinges and, as a consequence, the pier-pile system would have a higher horizontal capacity than 1207kN. It is very difficult to confirm this without accurate measurement of the excess pore water pressures generated during earthquake loading.

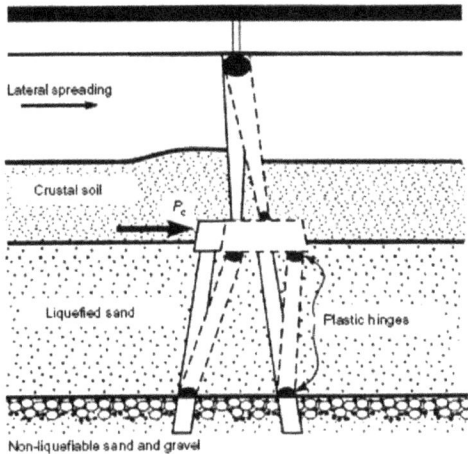

Fig. 1.14 Failure mechanism by formation of plastic hinges in the piles and in the pier.

From a performance point of view, the Landing bridge remained serviceable during the Edgecumbe earthquake, which can be taken as an L1 earthquake for this structure. Therefore, it can be assigned a performance grade of S or A following the nomenclature presented in Fig. 1.4.

This type of failure mechanism has been observed during many other earthquake events. For example, during the 921 Ji-Ji earthquake, a similar mechanism of lateral spreading ground is seen in Fig. 1.15. (Madabhushi, 2004). A newly constructed bridge pier in the vicinity can be seen in Fig. 1.16. Again, the passive soil bulge can be seen on one side and a gap is formed on the other side. In Fig. 1.16 it can be seen that the gap formed between the pier and the ground was filled up with liquefied soil that boiled up through the gap that was formed due to lateral spreading. Tokimatsu *et al.* (1996) present similar types of damage to pile foundations.

The failure mechanism shown in Fig. 1.14 is quite worrisome as the nonliquefied crustal soil layers can 'slide down' a gentle slope on a liquefied layer of sand. The crustal layer can be quite strong (for example, if it consists of over-consolidated clay layers) and therefore can generate large horizontal passive earth pressures on the piles and pile caps passing through it. In the limiting case a passive failure wedge can form causing the soil to bulge and heave as seen in Figs. 1.13 and 1.16.

Fig. 1.15 Lateral spreading next to a river channel during the 921 Ji-Ji earthquake, Taiwan.

Fig. 1.16 Passive bulge and gap formation next to a bridge pier following lateral spreading.

1.4.4 The Harbour Master's Tower at Kandla Port

Kandla Port is one of the largest ports in India and is located in the western state of Gujarat. Following the Bhuj earthquake of 2001, there was some damage to the port facilities. Kandla Port had a total of ten berths, which were a combination of old and new berths. In addition there were six oil terminals and a dry dock facility. The damage induced by the Bhuj earthquake was mainly to the port structures, with some storage facilities losing roof structures. (Madabhushi *et al.*, 2005). The old berths supported on battered piles performed reasonably well with only minor cracks appearing in the pile foundations as seen in Fig. 1.17a. Following the earthquake these berths were downgraded in terms of the loading that could be placed on them, while the cracks in the piles were investigated. The new berths were supported on steel-encased concrete pile foundations of about 1.1m diameter as shown in Fig. 1.17b. These pile foundations performed very well with no observable damage recorded during the earthquake. Liquefaction was observed in certain parts of the port and a subsidence of about 30cm was observed in the cargo area behind the new berths, as seen in Fig. 1.17c. A typical soil profile at Kandla Port was reported by Sitharam and Govindaraju (2004). The surface layer is made of soft plastic clay which extends to a depth of about 9m, as shown in Fig. 1.17d. Below this there is a substantial silty sand layer of about 13m thickness which is susceptible to liquefaction during strong shaking, as was produced by the Bhuj earthquake. This silty sand layer is underlain by reddish stiff clay of about 6m in thickness. Given this stratigraphy the pile foundations are generally taken into the stiff clay and are about 23–25m long. Also there is a good chance of the soft plastic clay layer at the surface suffering lateral spreading once the silty sand suffers liquefaction, thereby imposing lateral loads on the pile foundations.

The most interesting case at the Kandla Port is the Harbour Master's Tower (HMT) building located next to the old berths. This is a masonry building supported on 2×2 pile groups with each pile having a diameter of about 400mm and a pile length of 23m. Following the Bhuj earthquake, this building suffered differential settlements, leading to a rotation of about 11° to the vertical. A view of this building is shown in

Fig. 1.18, which shows detachment from the horizontal access ways on the left-hand side. However, despite being a masonry superstructure, very little cracking was seen.

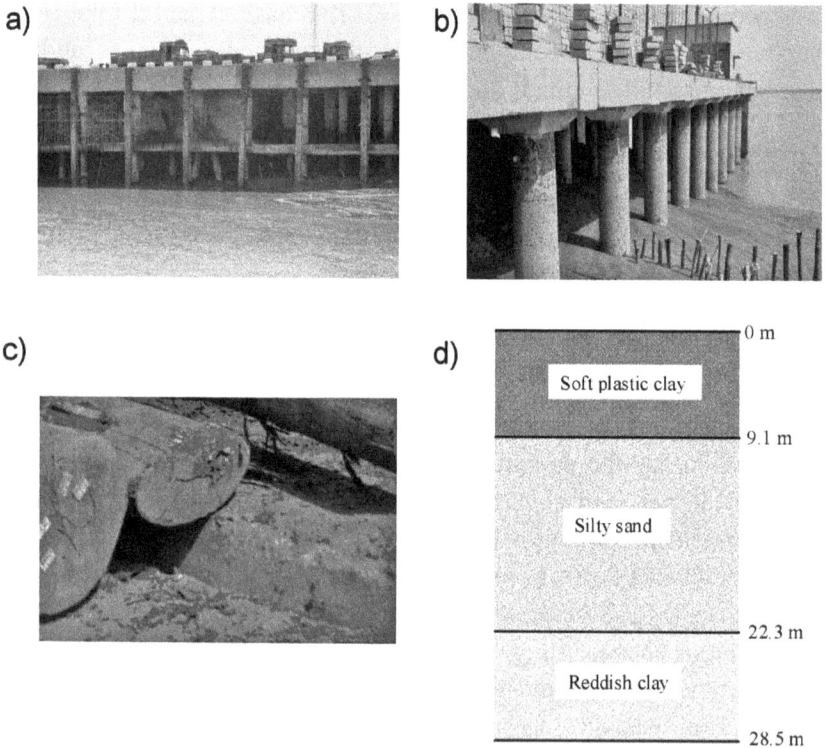

a)

b)

c)

d)

```
                                    0 m
   Soft plastic clay
                                    9.1 m

        Silty sand

                                    22.3 m
       Reddish clay

                                    28.5 m
```

Fig. 1.17 Liquefaction at Kandla Port, Gujarat, India: (a) view of the old berths with battered piles (b) view of the new berths with steel encased concrete piles (c) subsidence in the cargo handling area behind the new berths (d) typical soil profile at Kandla Port.

Liquefaction-induced lateral spreading was evident in the vicinity of the HMT building. A two-storey building next to the HMT building suffered a total collapse as shown in Fig. 1.19. This building lost the ground floor and appeared to have broken right down the middle. The foundations for this building were shallow, strip foundations. There was evidence of extensive liquefaction-induced lateral spreading and subsidence behind the two-storey building as well as the HMT building.

In Fig. 1.20 a view of the back of the HMT building towards the Gulf of Kachchh is presented, in which the extensive cracking to the parapet wall can clearly be seen. A trial pit was dug to expose the pile foundations at the right-hand bottom edge of the HMT building, as seen in Fig. 1.18. This confirmed that the settlement of the pile foundations had caused the rotation of the masonry superstructure.

Fig. 1.18 A view of the tilted HMT building.

Fig. 1.19 A view of the collapsed two-storey building.

Fig. 1.20 Lateral spreading caused the parapet walls behind the HMT building to crack severely.

The Bhuj earthquake of 2001 can be considered as an L2 event for the port facilities at Kandla, using the nomenclature introduced in Fig. 1.4. The performance of the old berths can be classed as Grade A due to some loss of usage in the period immediately after the earthquake. The new berths, on the other hand, can be classed as Grade S as they retained their full serviceability in the aftermath of the earthquake. The two-storey building had completely collapsed during this L2 event and therefore would be classified as Grade C. The HMT building would be classified between Grade A and Grade B as it suffered some damage during the L2 event but did not collapse fully.

From a pile design point of view the HMT building raises some interesting issues. Given the stratigraphy at Kandla Port (see Fig. 1.17d) and assuming that the Bhuj earthquake caused the silty sand to liquefy fully, it is reasonable to assume that the shaft friction of the pile was severely reduced during and immediately after the earthquake. The silty sand would also take a long time to recover its strength as excess pore pressures would take a long time to dissipate due to its relatively low permeability and the presence of the thick soft clay layer above. The axial load from the superstructure continues to act on the piles, while there is additional lateral loading on the piles and the pile cap due to the spreading of the soft clay. This raises the question of whether, under these circumstances, the pile would fail by forming plastic hinges and suffer a buckling instability or flexural bending failure as seen in the previous case of the NFCH building or would the piles simply try to mobilise additional base capacity by settling into the stiffer reddish clay layer below as seen in Fig. 1.17d?

1.5 Modes of Pile Failure in Liquefiable Soils

The four case histories described in the previous section and the performance of pile foundations in liquefiable soils listed in Table 1.4 form the basis of understanding the failure mechanisms suffered by pile foundations in liquefiable soils. In this section these failure mechanisms are considered first for single piles and later for pile groups. In the later chapters of this book, these failure mechanisms will be referred to, based on the experimental observations made in high-quality dynamic centrifuge tests or based on the results of analytical methods.

1.5.1 Failure mechanisms for single piles

Single piles with small pile caps are occasionally used to support individual columns of buildings. In this section the possible failure mechanisms of such single piles due to liquefaction of the ground will be presented. In certain circumstances a row of single piles may be used to support bridge piers as described in the case history of the Showa bridge failure (see Sec. 1.5.1). Under such circumstances those piles can suffer the same failure mechanisms. Firstly the case of single piles in level ground is considered. Two possible mechanisms of failure can be readily identified as shown in Fig. 1.21. In Fig. 1.21a, the single pile carrying large axial loads from the superstructure and located in a loose, liquefiable, saturated sandy layer overlying the bed rock is presented. When the earthquake-induced cyclic shear stresses lead to the generation of excess pore pressures in the sandy layer, the stiffness of this layer degrades significantly. Under those circumstances, the single pile can suffer buckling instability if sufficient length becomes 'unsupported' and can fail by forming a plastic hinge. In Fig. 1.21a the initial position of the pile is shown as a solid line and the anticipated failure mechanism is shown as a dotted line. Here, the location of the plastic hinge shown as a solid circle in Fig. 1.21a is of interest. It is reasonable to assume that the plastic hinge will form close to the interface of the liquefied sandy layer and the bed rock as shown in Fig. 1.21a. However, experimental evidence will be presented in Chapter 3, in which the plastic hinge actually forms some way above the location shown in Fig. 1.21a, indicating that the liquefied sand can generate some 'resistance' to the buckling pile.

In contrast to this, the case of a single pile that carries large axial loads and passes through the loose, saturated layer and rests in a dense sand layer is presented in Fig. 1.21b. The loose, sandy layer will again see a significant rise in excess pore pressures and the consequential degradation of the soil stiffness. However, the dense sand layer will also see a significant increase in pore pressure. Coelho *et al.* (2003) have reported the generation of excess pore pressures in dense sand deposits. Further, the excess pore pressure generated in the loose sand layer close to the interface between the loose and dense sand layers can be transmitted into the dense sand layer, further softening the dense sand. These conditions will lead to

- loss of shaft friction of the pile; and
- a reduction in the base capacity of the pile.

Under these circumstances the pile foundation will suffer a bearing failure and will settle into the dense sand layer as shown in Fig. 1.21b. Settlement will continue until sufficient base capacity and shaft friction is mobilised to bring the axial load and the weight of the pile into equilibrium. The dilation of the dense sand layer will generate large suction pressures in the dense sand, which will also aid the process of reaching a vertical equilibrium in the short term. (Coelho *et al.*, 2003 and 2007).

a)   b)

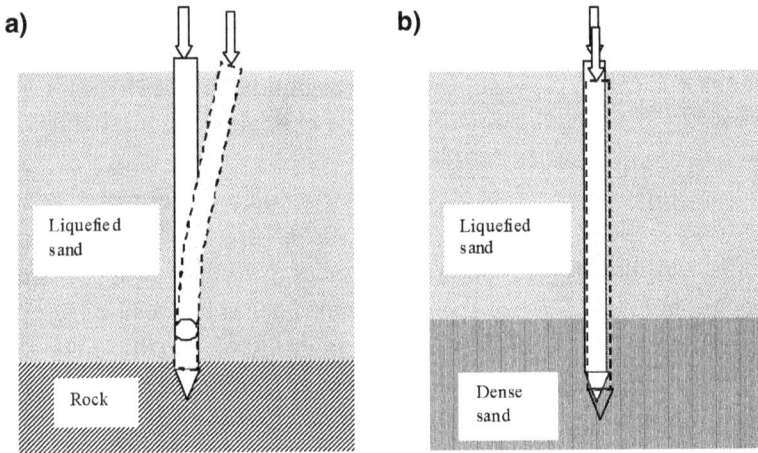

Fig. 1.21 Modes of collapse for single piles in liquefiable soil (a) buckling instability (b) bearing failure.

In many of the cases of pile performance referred to in Table 1.4 and those described in Sec. 1.5 the poor pile performance was also accompanied by lateral spreading of the ground. It is therefore necessary to consider the possible failure mechanisms of pile foundations in the presence of lateral spreading. Haigh *et al.* (2000) have established that even gently sloping ground of 2° or 3° to the horizontal can suffer lateral spreading. So the field cases where lateral spreading is to be anticipated will be rather large, increasing the importance of this problem.

In Fig. 1.22a a single pile that is located in an inclined, liquefiable sandy layer is shown. The pile carries significant axial load and is rock-socketed into the bed rock. Following earthquake-induced liquefaction, the sandy layer will suffer lateral spreading from left to right in Fig. 1.22a. The single pile can suffer a buckling instability similar to the one shown in Fig. 1.21a if a sufficient length of the pile becomes unsupported, only more easily than in level ground owing to presence of a lateral driving force offered by the lateral spreading. Further, the critical buckling load estimated using Euler's theory will reduce in the presence of a lateral driving force on the pile. A further variation on this type of failure mechanism is the presence of a nonliquefied soil layer that overlies the liquefied layer. Such a nonliquefied crustal layer can suffer lateral spreading moving down-slope often associated with the formation of a thin film of water at the interface between the liquefied and nonliquefied layers, especially when the nonliquefied layer has a lower permeability than the liquefied layer. The pile will be subjected to large, passive, lateral pressures by the nonliquefied crust. In the presence of axial load, P-δ effects can become excessive and together with the lateral loads can cause formation of plastic hinges as shown in Fig. 1.22b. Although the final failed shape of this pile may appear similar to the one in Fig. 1.22a, the specific circumstances here cause the pile to fail by flexural bending rather than buckling. Also, the thickness of the liquefiable layer can be relatively small for this type of failure mechanism, as buckling is not the issue.

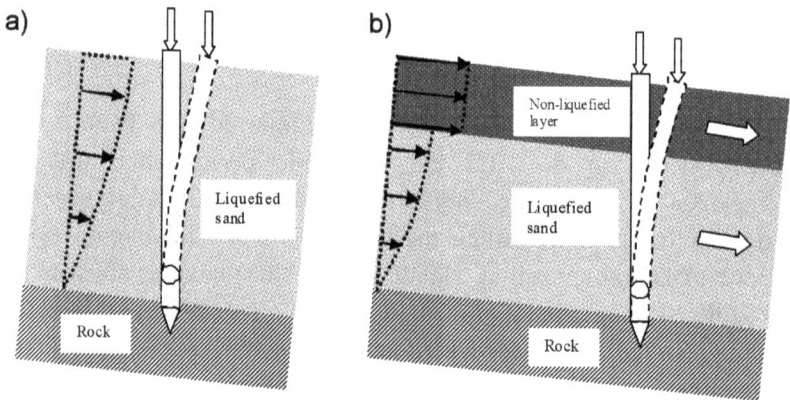

Fig. 1.22 Failure of piles under combined lateral and axial loads in laterally spreading soil (a) liquefiable sand only (b) with nonliquefied crustal layer.

A final variation of the commonly occurring field cases is when the pile is located in layered ground with a nonliquefied crust underlain by a liquefiable layer that is in turn underlain by a dense sand layer or a stiff clay layer. The pile will normally be driven to come to bear in the dense sand layer to transfer the axial loads. Following liquefaction and lateral spreading, the pile can suffer some elastic bending due to the lateral earth pressure exerted by the nonliquefied layer. However, the pile may concurrently suffer bearing failure as shown in Fig. 1.23. This is a variation on the failure mechanism in level ground presented in Fig. 1.21b.

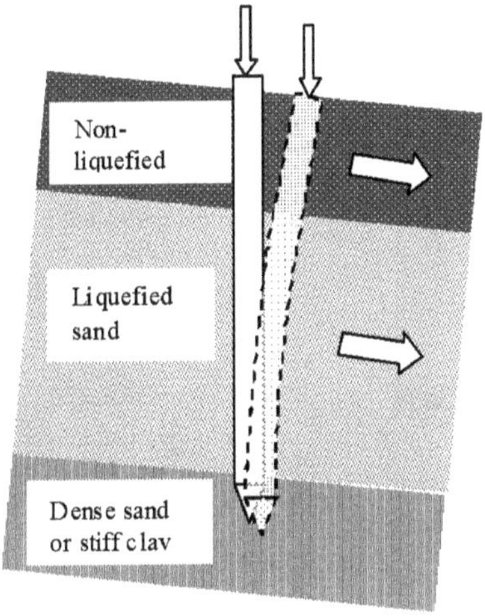

Fig. 1.23 Combined bending and settlement failure of a pile in laterally spreading ground.

1.5.2 Failure mechanisms for pile groups

Deep foundations are more commonly designed with groups of piles. The failure mechanisms for pile groups in level ground are considered next. The case of a single liquefiable layer overlying the bedrock is considered. Again, axial load is present on the pile group at the time of the earthquake loading. As described before, the liquefiable sand layer will lose its stiffness owing to the generation of excess pore water pressures. Under those circumstances, the piles become 'unsupported' and can fail as before due to buckling instability. Two further mechanisms of failure are possible as presented in Fig. 1.24. In Fig. 1.24a the piles fail by forming plastic hinges both at the bases and the pile heads. This assumes that the piles are well rock-socketed into the bedrock. If this is not the case, i.e. the pile tips rest on the bedrock, then no plastic hinges will form at the pile tips. In this mechanism the pile cap will suffer lateral displacements and small vertical settlements that may cause significant distress to the superstructure.

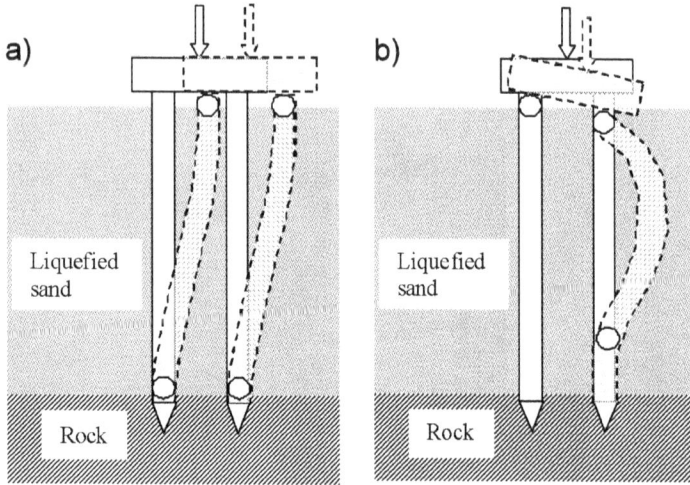

Fig. 1.24 Pile group instability in level ground (a) four hinge mechanism (b) three hinge mechanism.

A second mechanism that can occur is a three-hinged mechanism shown in Fig. 1.24b. In this case the plastic hinges will form only at the pile heads. One of the piles has to fail by buckling instability. This mechanism can cause severe rotation of the pile cap causing significant distress to the superstructure. Also note that this mechanism can occur even if the pile tips are well rock-socketed to mobilise full fixity at the pile tips.

As before, the pile performance in liquefiable soils is more interesting in the presence of laterally spreading soils. The failure mechanisms for these cases are considered next.

The soil profile considered is the same as before with a nonliquefied crust layer underlain by a liquefiable sand layer, which in turn is underlain by bedrock. The pile group is socketed into the rock. The axial load from the superstructure is present on the pile group at the time of the earthquake loading. The failure mechanism for the pile group is shown in Fig. 1.25. Large lateral loads will be generated on the piles due to the passive earth pressures from the nonliquefied crust on the piles and pile cap. This, in combination with the axial load, can cause excessive

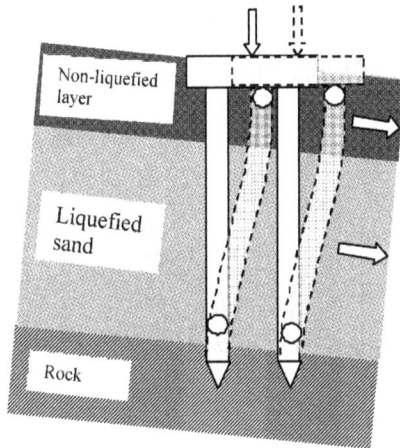

Fig. 1.25 Pile group failure in laterally spreading ground with nonliquefied crust.

bending of the piles leading to formation of four plastic hinges as shown in Fig. 1.25, similar to the one for level ground shown in Fig. 1.24. In laterally spreading ground it is easier to form this mechanism than in level ground, due to the presence of large lateral loads.

When the tips of a pile group rest in a layer of dense sand or stiff clay, the pile group can additionally suffer settlements as shown in Fig. 1.26. The lateral loads from the nonliquefied crust will cause some elastic bending of the piles. However, the pile group can fail by loss of end-bearing and therefore suffer settlements. In the case of the pile group, this mechanism of failure is somewhat limited by the ability of the pile cap to resist settlements as it rests in the nonliquefied crust layer. Unlike the single pile shown in Fig. 1.23, this additional bearing capacity of the pile cap can limit excessive settlements of the pile group.

In a similar type of soil stratification, further failure mechanisms are possible for the pile group as shown in Fig. 1.26. One of the piles can suffer excessive settlement as shown in Fig. 1.26b. This settlement can be accommodated by the formation of plastic hinges at the pile cap as shown in Fig. 1.26b. This failure mechanism can lead to excessive rotation of the pile cap. This type of failure mechanism would be compatible with the HMT building rotation seen in the case history of Kandla Port (see Sec. 1.4.4).

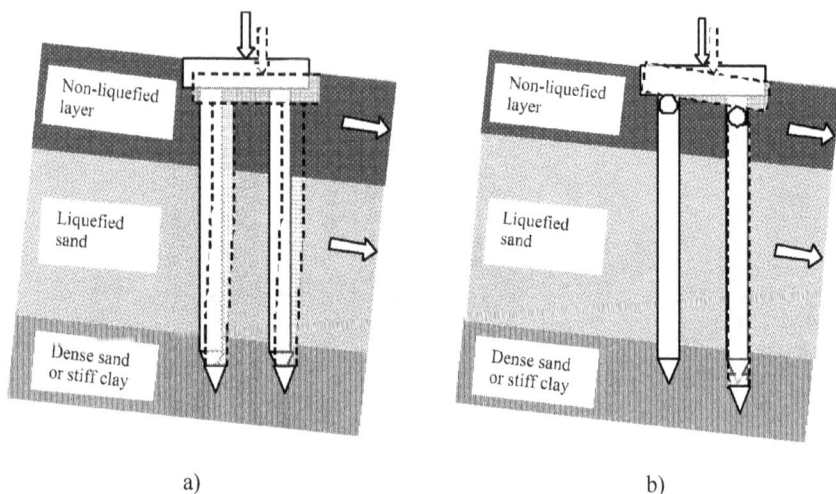

a)   b)

Fig. 1.26 Bearing failure of pile groups in laterally spreading ground (a) bearing failure alone (b) combination of local bearing failure and plastic hinging.

1.6 Summary

Pile foundations are the most popular form of deep foundations. In this chapter the design of pile foundations subjected to axial loading was briefly considered and the two main components of load transfer, i.e. the shaft friction and base capacity, were discussed. Earthquakes can create a particularly challenging set of loads on piles, especially when the soil layers in the vicinity of the foundations suffer liquefaction. Lateral spreading of sloping ground can additionally subject the piles to large, horizontal loads. In this chapter some of the case histories of the performance of the piled foundations in ground that suffered liquefaction were considered. Also the performance of these foundations was classified based on the modern approach of performance grades in L1 and L2 earthquake events. Based on the observations of failures and performances in these case histories, a set of possible failure mechanisms are described for single piles and pile groups in liquefied level ground and laterally spreading sloping ground.

Chapter 2

Inertial and Kinematic Loading

2.1 Pile Behaviour Under Earthquake Loading

For many classes of structure the predominant static loading on piled foundations is vertical compressive loading. Earthquake loading will impose requirements on the piles to resist significant lateral loads and moments with the further possibility of piles being required to carry tensile loads. The deformation of piles may be substantially affected by the permanent deformations of the ground in which they are embedded. In particular, liquefaction induced lateral spreading can impose severe damage on piled foundations as discussed in Chapter 1.

Pile foundations can suffer severe damage during, and in the period immediately after, the earthquake loading. In the previous chapter several case histories of pile performance during past earthquakes were discussed. Based on the case histories and the state of current research, some of the possible failure mechanisms of pile foundations in various soil stratigraphies were presented. In this chapter, the actions of the earthquake loading on the pile foundations are considered in detail. These are usually classified as *inertial* and *kinematic* loadings depending on the whether the action on the pile is induced by the superstructure or the soil surrounding the pile.

2.1.1 Inertial loading

The loading requirements imposed by seismic events on piles require different geotechnical and structural design of these elements compared with the static design of piles under static loading. Let us first consider the case of pile foundations in level ground comprising a soft, horizontal soil layer in which no soil liquefaction occurs, overlying a stiff, horizontal soil layer. The pile foundation passes through the soft layer and rests on the stiff soil layer as shown in Fig. 2.1.

The earthquake motion can be transmitted from the stiff soil stratum into the softer layer and this motion can be amplified as it propagates through the softer layer, is transferred to the piles and onto the superstructure. This sets up structural vibrations in the superstructure. As these vibrations are being set up in the superstructure, it will impose inertial loading on the pile cap. This inertial load has to be carried by the piles. The peak inertial load is shown schematically in Fig. 2.1. If the inertial load is large, then piles can suffer significant lateral displacements. Further, depending on the stiffness of the superstructure and the pile cap's bearing capacity, the pile cap can either be prevented from suffering any rotation or some rotation can occur. If the pile cap rotations are prevented then the piles will undergo elastic bending as shown in Fig.1 and in extreme cases plastic hinges can form at the pile/pile cap interface as indicated in Fig. 2.1. On the other hand, if the pile cap can undergo some rotation, then the piles will suffer elastic bending but the pattern of bending is opposite to the previous case (see Fig. 2.1). These types of behaviour will be dominated by the stiffness of the shallow soil layers as the lateral inertial load on the piles will only displace the soil in this region. If the shallow layers in this region are soft, this can be accomplished easily. On the other hand, if the shallow layers below the pile cap comprise stiffer/competent soil then the lateral displacement of piles and the rotation of the pile cap will be smaller, but the piles may attract large lateral loads that oppose the inertial loads due to mobilisation of the passive soil pressures in this region.

Clearly 'inertial loading' on the pile is a dynamic load and its peak magnitude would depend on the dynamic characteristics of both piles and the soil layers. In particular the inertial load will depend on the stiffness of the pile relative to that of the surrounding soil. This aspect is further considered in Sec. 2.6. In addition, the dynamic response of the superstructure will play an important role in the magnitude of inertial load seen by the piles.

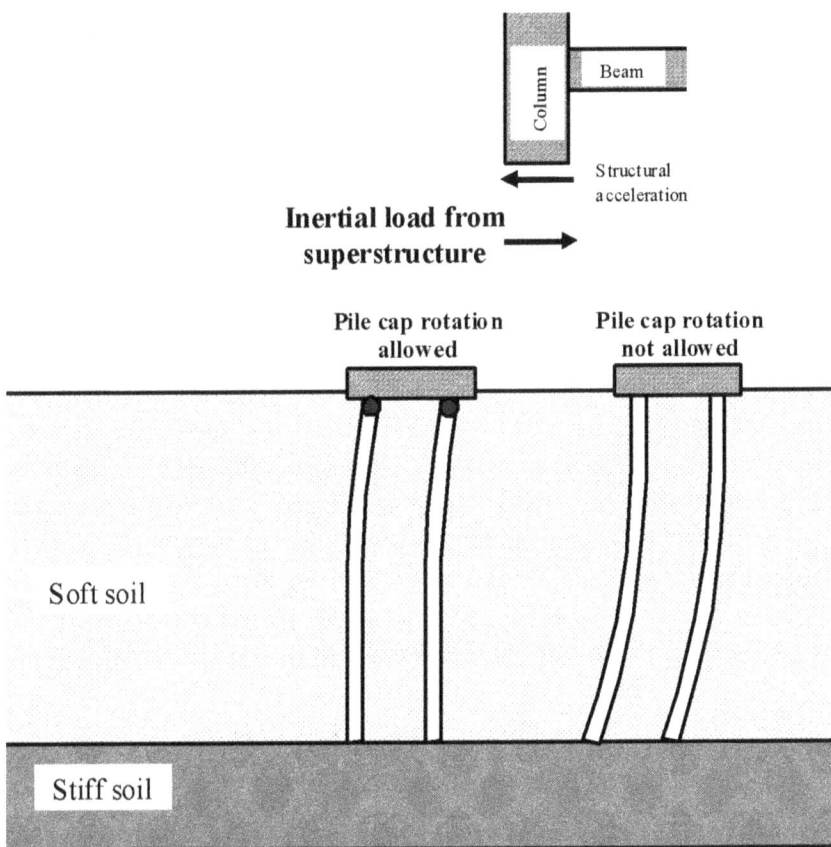

Fig. 2.1 Inertial interaction between piles and the soil.

Further, the inertial loading on the piles would be the same in both the level ground scenario as well as in sloping ground that may be suffering lateral spreading, provided that the changes in the dynamic response of the piles due to the liquefaction-induced softening are small.

2.1.2 Kinematic loading in level ground

Earthquake loading differs from other forms of environmental and machinery-induced cyclic loading because the in-ground motions produce pile loadings in addition to the pile loadings derived from the motion of the supported structure. The in-ground motion generates 'kinematic interaction' between the piles and the soil. As the soft soil layer undergoes vibrations due to the arrival of cyclic shear stresses from the stiffer layer below, it will vibrate in the first mode, with a natural frequency of f_n;

$$f_n = \frac{v_s}{4H} \qquad (2.1)$$

where v_s is the shear wave velocity in the soft layer, and H is the thickness of the soft layer. This motion is indicated as free-field soil displacement in Fig. 2.2. The piles that are located in such a soil layer are forced to follow the motion imposed by the soil. If the piles are relatively flexible then they will undergo deformations as shown schematically in Fig. 2.2. On the other hand if the piles are relatively stiff, then the deformations they undergo will be smaller. In this way, the piles can attract large lateral loads that will act in the opposite direction to the imposed displacements, due to mobilisation of passive pressures.

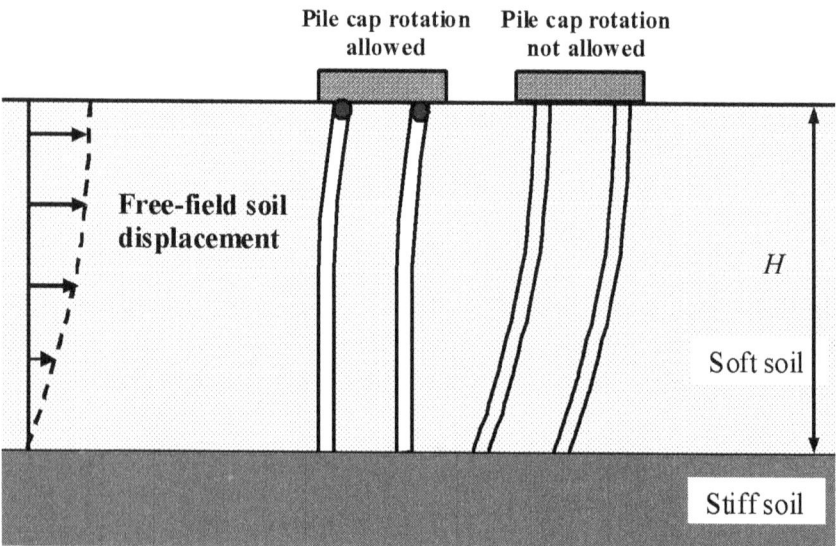

Fig. 2.2 Kinematic interaction between the piles and soil.

It is clear that due to the kinematic interaction between the soil and the pile, the deformations in the piles occur in the deeper regions as indicated in Fig. 2.2. The variation of inertial and kinematic loading on piles with increasing depth of soil layers is considered later in Sec. 2.3.1.

2.1.3 Kinematic loading in sloping ground

The kinematic loading on the piles can increase in sloping ground. Earthquake loading can cause down-slope movements to occur when the cyclic shear stresses generated by the earthquake combined with the static shear stresses in the slope exceed the shear strength of the soil. In such situations the free field displacements (as shown in Fig. 2.2 for level ground) will be larger and therefore the kinematic loading on the piles will be larger. If the sloping ground suffers liquefaction-induced lateral spreading as seen in the case histories in Chapter 1, again the free-field displacements in the sloping ground will be larger. However, the kinematic loading may not be large as the strength of the liquefied soil will be low and as such it may be able to flow around the piles. There are exceptions to this situation. For example, if there is a nonliquefied crust above the laterally spreading liquefied layer as discussed in Sec. 1.4 in Chapter 1, then the kinematic loading on the piles can be quite large. The nonliquefied layers can result in quite a large kinematic loading as the piles and the pile cap can be subjected to large passive pressures due to the higher strength of this soil. Before presenting the methodologies for estimating inertial and kinematic loads on pile foundations, the development in the methods of analysis of static lateral loads on piles is considered first.

2.2 Analysis of Laterally Loaded Piles Under Static Conditions

Pile foundations can be subjected to lateral loads under static conditions in a variety of loading scenarios. For example, pile foundations supporting a bridge abutment may be subject to lateral loads due to the earth pressures acting on the abutment. The static load-deflection analysis of piles has developed in two principle directions. These methods are: the Winkler spring approach in which the pile is modelled as a beam supported by a series of independent springs; and the elastic continuum approach in which an elastic pile is considered to be embedded in an elastic soil continuum. These two approaches can differ from the realistic situations faced in the field as illustrated in Fig. 2.3.

The different approaches have different strengths and weaknesses. The Winkler spring method allows the nonlinear loading response of the soil to pile deflection to be easily incorporated through the use of nonlinear p-y or t-z curves. These springs can be modified to incorporate the effects of imposed ground movements around the piles. In addition, complex layered soil profiles can also be accommodated in a straightforward manner. However, the springs do not account for the effects of soil movement at one location on soil movements at adjacent locations. This limits the reliability of the empirical methods used and makes the analysis of pile groups difficult with this method.

On the other hand, the elastic continuum approach is more satisfactory from a theoretical standpoint as the stress and strain fields in the soils around the pile are correctly analysed. This makes the technique suitable for the analysis of the interaction of piles in pile groups. However, the available solutions are predominantly linear-elastic and based on rather simple soil profiles.

2.2.1 Simplified soil profiles

For the purposes of carrying out simple analyses of piles subjected to lateral loads either using the p-y method or in some cases even finite element analysis with elastic constitutive relationships for the soils, the soil layers around the pile are assumed to have certain simple variations in shear stiffness. The most common variations considered in the literature are presented in Fig. 2.4. In this figure it can be seen that the shear stiffness is either assumed constant with depth (Fig. 2.4a), which is applicable in the case of over-consolidated clays. For sandy soils it is more appropriate to assume a parabolic variation in stiffness with depth as shown in Fig. 2.4b. This is akin to the variation of shear modulus proposed by Hardin and Drnevich (1972) for sandy soils and expressed in SI units as shown in Eq. 2.2.

$$G_o = 100 \frac{(3-e)^2}{(1+e)} \sqrt{p'} \qquad (2.2)$$

where G_o is small-strain shear modulus in MPa, e is the void ratio of the sand and p' is the mean effective confining stress in MPa.

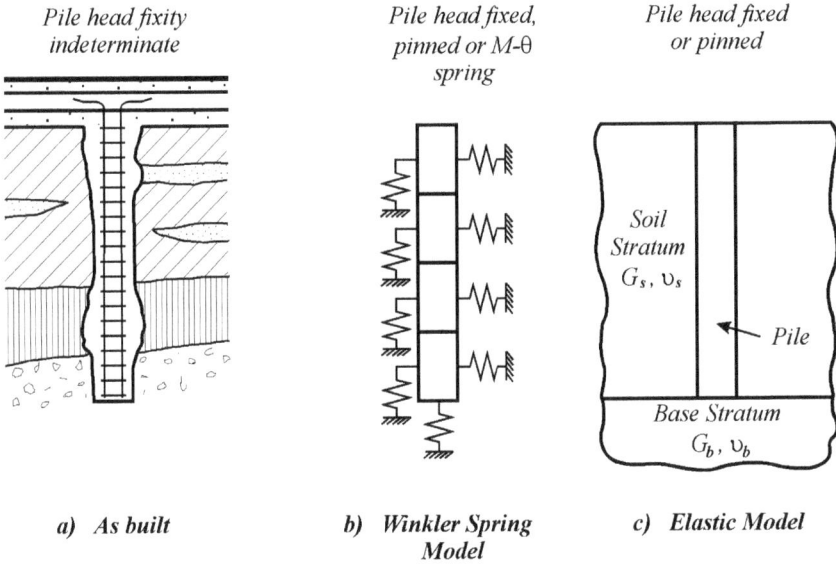

a) As built   *b) Winkler Spring Model*   *c) Elastic Model*

Fig. 2.3 Alternative models for pile load–deflection analyses.

For soft clay layers it may be more appropriate to assume a linear variation in stiffness with depth as shown in Fig. 2.4c. The advantage of assuming these kinds of simplified variations in stiffness is that they can be easily incorporated into the calculation of spring stiffness in the horizontal and vertical direction for *p-y* type and *t-z* type analyses respectively. Similarly, if continuum based finite element analyses are being attempted again these type of variations in soil stiffness are amenable to straightforward implementation. It must be pointed out that these simplified variations can only be considered when the soil layers are homogeneous. When the soil strata under consideration have distinct layers, suitable approximations have to be made.

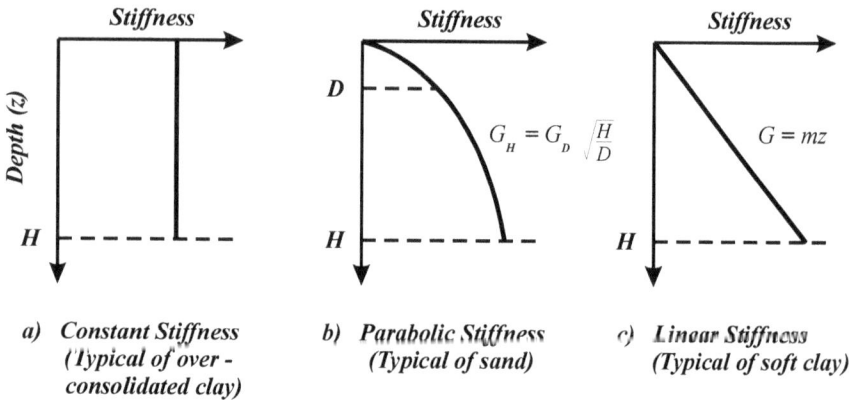

a) *Constant Stiffness*
    (Typical of over -
    consolidated clay)

b) *Parabolic Stiffness*
    (Typical of sand)

c) *Linear Stiffness*
    (Typical of soft clay)

Fig. 2.4 Idealised soil stiffness profiles.

2.2.2 Simplified analysis procedures for piles under static loading

Solutions for single piles under static lateral loading are given by Poulos and Davis (1980) with further solutions by Davies and Budhu (1986), Budhu and Davies (1987, 1988) and Gazetas (1990, 1991). These are summarised by Pender (1993) for the Winkler spring and elastic models for a variety of stiffness distributions. The strengths of both methodologies can be harnessed by using the Winkler spring model to refine the soil stiffnesses selected for horizontal and vertical elastic analysis of single piles. The refined parameters from the single pile analysis may then be employed in an elastic-continuum analysis of the pile group.

For static lateral loading of an elastic pile embedded in an elastic soil, the displacement u and rotation θ of the pile head are given by:

$$\Delta = f_{uH}H + f_{uM}M \qquad (2.3)$$

and

$$\theta = f_{\theta H}H + f_{\theta M}M \qquad (2.4)$$

where H is the horizontal load and M the moment at the pile head, and f_{uH}, f_{uM}, $f_{\theta H}$, $f_{\theta M}$ are flexibility coefficients with $f_{\theta H} = f_{uM}$. The pile head flexibility coefficients for the three soil stiffness profiles given in Fig. 2.4 may be expressed as shown in Table 2.1.

Table 2.1 Pile head flexibility coefficients for static loading.

Flexibility Coefficient	Soil stiffness variation		
	Constant	Parabolic	Linear
f_{uH}	$\dfrac{1.3}{E_s D_0}\left(\dfrac{E_p}{E_{sD}}\right)^{-0.18}$	$\dfrac{2.14}{E_{sD} D_0}\left(\dfrac{E_p}{E_{sD}}\right)^{-0.29}$	$\dfrac{3.2}{m D_0^{\,2}}\left(\dfrac{E_p}{E_{sD}}\right)^{-0.333}$
$f_{\theta H} = f_{uM}$	$\dfrac{2.2}{E_s D_0^{\,2}}\left(\dfrac{E_p}{E_{sD}}\right)^{-0.45}$	$\dfrac{3.43}{E_{sD} D_0^{\,2}}\left(\dfrac{E_p}{E_{sD}}\right)^{-0.53}$	$\dfrac{5.0}{m D_0^{\,3}}\left(\dfrac{E_p}{E_{sD}}\right)^{-0.556}$
$f_{\theta M}$	$\dfrac{9.2}{E_s D_0^{\,3}}\left(\dfrac{E_p}{E_{sD}}\right)^{-0.73}$	$\dfrac{12.16}{E_{sD} D_0^{\,3}}\left(\dfrac{E_p}{E_{sD}}\right)^{-0.77}$	$\dfrac{13.6}{m D_0^{\,4}}\left(\dfrac{E_p}{E_{sD}}\right)^{-0.778}$

where m = rate of increase of stiffness with depth.

The matrix of pile head flexibility coefficients can be inverted to obtain the matrix of pile head stiffness coefficients K_{HH}, K_{HM}, K_{MH} and K_{MM} as shown in Eq. 2.5. These can be employed to define horizontal and rotational springs, which reproduce the pile head displacement and rotation under a given static horizontal and moment load.

$$\begin{bmatrix} K_{HH} & K_{HM} \\ K_{MH} & K_{MM} \end{bmatrix} = \frac{1}{\left(f_{uH} f_{\theta M} - f_{uM}^2\right)} \begin{bmatrix} f_{\theta M} & -f_{uM} \\ -f_{\theta H} & f_{uH} \end{bmatrix} \tag{2.5}$$

where $K_{HM} = K_{MH}$, giving a symmetric stiffness matrix. The overall horizontal pile head stiffness (K_h) is therefore given by

$$K_h = \frac{K_{HH} K_{MM} - K_{HM}^2}{K_{MM} - e K_{HM}} \tag{2.6}$$

where e is the eccentricity given by $e = M / H$. K_h will have units of kN/m if the horizontal load H and moment load M are in kN and the eccentricity e is in m. Similarly, the rotational stiffness at the pile head (K_θ) is given by

$$K_\theta = \frac{K_{HH}K_{MM} - K_{HM}^2}{K_{HH} - K_{HM}/e}$$ (2.7)

K_θ will have units of kN-m/rad.

2.3 Analysis of Laterally Loaded Piles Under Earthquake Loading

2.3.1 Variation in the action of inertial and kinematic loads with depth

Summaries of the methods used to assess the responses of piles and pile groups to seismic loading are provided by Gazetas (1984), Novak (1991) and Pender (1993). Numerical studies indicate that the response of a pile shaft under seismic loading can be considered in three zones:

i.   The *near surface zone*. This zone extends to approximately eight pile diameters beneath the soil surface and is dominated by inertial loading effects.

ii. An *intermediate zone*. This zone exists between the near surface and deep zones and is influenced by both inertial and kinematic effects.

iii. The *deep zone*. This zone is below 12 to 15 pile diameters from the surface and is dominated by kinematic effects.

It is interesting to consider whether the peak inertial and kinematic loads in the intermediate zone occur simultaneously and therefore must be superimposed or if the kinematic effects only occur after the end of peak earthquake loading, i.e. after the inertial loading has been completed. Further distinction needs to be made on whether the surrounding soil will or will not liquefy. If the soil liquefies then the kinematic loading may be small as the strength of liquefied soils may be small by conventional wisdom, although experimental evidence suggests that this may not be the case due to dilation of liquefied soil during shear deformation. This aspect will be discussed in detail in Chapter 4. Similarly another distinction needs to be made between liquefied level ground and liquefied sloping ground that is undergoing lateral spreading, say with a nonliquefied crust on the top as discussed in Chapter 1. The peak kinematic loading applied by such laterally spreading nonliquefied crusts may occur after the passage of peak inertial loads applied by the superstructure. Therefore, such kinematic loads may not be superimposed over the peak inertial loads. Recent research by Davis at the University of California has investigated this aspect for a pile group in laterally spreading ground with a nonliquefied crust on the surface. (Brandenberg *et al.*, 2005a) present an interesting discussion of the phase relationship between the peak inertial and kinematic loads before and during lateral spreading.

2.3.2 Effective lengths of piles

The effective length of pile L_{ad} that participates in the inertial response may be determined for elastic soil profiles as a function of the relative stiffness of the pile with respect to the surrounding soil and the pile diameter. (Gazetas, 1984). With reference to the idealised soil profiles shown in Fig. 2.4, the effective length of pile L_{ad} may be calculated using the following equations.

Constant stiffness with depth:

$$L_{ad} = 2D_0 \left(\frac{E_p}{E_{sD}} \right)^{0.25}$$ (2.8)

Parabolic stiffness with depth:

$$L_{ad} = 2D_0 \left(\frac{E_p}{E_{sD}} \right)^{0.22}$$ (2.9)

Linearly increasing stiffness:

$$L_{ad} = 2D_0 \left(\frac{E_p}{E_{sD}} \right)^{0.20}$$ (2.10)

where, D_0 is the diameter of the pile, E_p is the Young's Modulus of pile and E_{sD} is the Young's Modulus of soil at a depth D_0. It must be pointed out that the use of Young's modulus of soil as a parameter is rather unsatisfactory as it is not easy to determine from element tests and practicing engineers often have no feel for this parameter.

The effective lengths calculated using Eqs. 2.8 to 2.10 are somewhat greater than the equivalent lengths which can be determined for piles under static loading. Field studies such as those by Hall (1984) and Makris *et al.*, (1996) on instrumented piled structures under significant levels of seismic loading show that the stiffness of a pile group tends to decrease significantly as the number of load cycles increases. This is due to effects such as a decrease in soil stiffness as shear induced pore water pressures increase in the near surface zone and the development of gapping around the top of the pile shafts. These effects will increase the effective or active length of the piles to be considered in the inertial loading response.

The effective pile length concept is useful for differentiating between 'long' and 'short' piles. For 'long' piles an increase in length does not affect the horizontal response to inertial loading. 'Short' piles, of length less than L_{ad}, exhibit a softer response to inertial load, which is a function of pile length.

2.3.3 Pile flexibility

As an alternative to the method suggested above the flexibility of the pile can be determined using the following procedure. The elastic length of pile can be determined using

$$T = \left(\frac{E_p I_p}{k} \right)^{0.2} \qquad (2.11)$$

where $E_p I_p$ is the flexural stiffness of the pile and, k is gradient of the soil modulus which may vary from 200 to 2000 kN/m^3. The value of k is 2000 kN/m^3 for loose saturated conditions. Thus elastic length of the pile is determined as a function of the relative pile–soil stiffness.

Using the value of T calculated above, Z_{max} is calculated using Eq. 2.12:

$$Z_{max} = \frac{L}{T} \qquad (2.12)$$

If $Z_{max} > 5$, the pile is considered to be *flexible*. i.e. its behaviour is not affected by the length, and collapse is always caused by a flexural failure, with formation of a plastic hinge. The pile is *semi-flexible* if $5 > Z_{max} > 2.5$, and the pile is considered *rigid* if $Z_{max} < 2.5$.

Piles that are classified as flexible will 'move' with the surrounding soil and therefore will attract the inertial shear load imposed by the superstructure during earthquake loading, but will attract little kinematic loading. Rigid piles, on the other hand, will attract significant kinematic load, as the piles stay in position so that the soil exerts passive pressures on either side of the pile in alternative load cycles due to the relative soil-pile displacement. This additional lateral load applied by the soil must be considered in pile design.

2.4 Kinematic Response in Level Ground

It is convenient to analyse the kinematic response of the pile or pile group separately from the inertial response. The kinematic response at depth may be used to assess the structural requirement of the pile in the intermediate and deep zones. The kinematic response of the pile head is an input into the inertial response analysis.

In the deep zone the presence of piles has little effect on the ground motion or natural frequency of the stratum as determined from Eq. 2.1. The pile and soil motions are likely to be practically coincident for frequencies up to at least 1.5 times the natural frequency (f_n) of the stratum. This observation is of practical significance as the deflected shape of the pile can be obtained from a 1D equivalent linear shear wave propagation analysis. Having obtained the deflected shape of the pile, its bending moments and shear forces may readily be determined. Makris *et al.* (1996) discuss field studies which make useful observations on this mode of behaviour. It should be noted that substantial bending moments may be induced in piles at the levels of interfaces between zones of appreciably different stiffness.

In order to perform an inertial response analysis the kinematic pile head response is required. Numerical studies indicate that the kinematic response derived for a single pile is applicable to pile groups and that the kinematic interaction between the soil and a free headed pile is conservative if applied to a fixed head pile.

Pender (1993) describes an approximate technique based on Gazetas (1984), which may be used to evaluate the kinematic response of the pile head. Firstly, the free field response at the top of the soil column is determined at a point remote from the pile group. The horizontal amplitude of the free field motion is u_o. Then a frequency dependent horizontal interaction factor I_u is determined using Eq. 2.13.

$$I_u = \frac{u_p}{u_o} \tag{2.13}$$

where u_p – horizontal amplitude of the pile head motion. In addition to the horizontal interaction factor I_u, a dimensionless factor F is calculated using Eqs. 2.14 to 2.16 for the three types of stiffness variation presented in Fig. 4.

Constant stiffness with depth:

$$F = \left(\frac{f}{f_n}\right)\left(\frac{E_p}{E_{sD}}\right)^{0.30}\left(\frac{L}{D}\right)^{-0.5} \tag{2.14}$$

Parabolic variation in stiffness with depth:

$$F = \left(\frac{f}{f_n}\right)\left(\frac{E_p}{E_{sD}}\right)^{0.16}\left(\frac{L}{D}\right)^{-0.35} \tag{2.15}$$

Linearly increasing stiffness:

$$F = \left(\frac{f}{f_n}\right)\left(\frac{E_p}{E_{sD}}\right)^{0.10}\left(\frac{L}{D}\right)^{-0.4} \tag{2.16}$$

where f is the response spectrum frequency considered, f_n is the natural frequency of the soil stratum, E_p is the Young's modulus of the pile, E_{sD} is the Young's modulus of soil at depth D, and L and D are the length and diameter of the pile, respectively.

Using the appropriate equation, values of F are calculated for discrete frequencies across the frequency range of interest (e.g. 0.5Hz to 40Hz). Corresponding values of I_u are calculated from the following expression (Gazetas, 1984):

$$I_u = aF^4 + bF^3 + cF^2 + 1.0 \qquad (2.16)$$

While using Eq. 2.16, I_u is limited to a minimum value of 0.5. The coefficients a, b and c in Eq. 2.16 are given in Table 2.2.

Table 2.2 Coefficients for horizontal kinematic interaction factor

Coefficient	Soil Stiffness Profile		
	Constant	Parabolic	Linear
a	0	3.64×10^{-6}	-6.75×10^{-5}
b	0	-4.36×10^{-4}	-7.0×10^{-3}
c	-0.21	6.0×10^{-3}	3.3×10^{-2}

The interaction factors produced by this procedure are strictly applicable only to a Fourier spectrum. However, approximate results can be obtained by applying the interaction factors directly to the free field spectral acceleration versus frequency response spectrum. The horizontal spectral acceleration of the pile head is simply obtained by multiplying the free field acceleration by the value of I_u for each frequency considered.

Study of spectral acceleration responses produced by the above procedure shows that the piles damp the higher frequency excitation seen in the free field. The extent of that damping depends on the ratio of pile to soil stiffness and particularly on the soil stiffness profile. The linearly increasing stiffness profile produces damping at lower frequencies than the other stiffness profiles. Pender (1993) observes that the response of instrumented piles in earthquakes tends to that of the linearly increasing stiffness profile even if the SI data suggest a constant or parabolic profile. This is considered to be due to softening of the soil with shear strains close to the surface, under the action of seismic loading.

The studies undertaken by Gazetas (1984) show that the rotational interaction factor is sufficiently small to be neglected.

2.5 Kinematic Loading in Laterally Spreading Soil

As described in Chapter 1, a more serious lateral loading condition on piles may arise when there is a nonliquefied crust situated above a liquefiable soil deposit on sloping ground. Such a condition will invariably result in lateral spreading following an earthquake event, thereby inducing large lateral loads on the piles and pile cap. Recent research at the University of Cambridge, UK (Haigh and Madabhushi, 2005), Renesselaer Polytechnic Institute, New York, US (Dobry *et al.*, 2003) and University of California, US, Davis (Chang *et al.*, 2005) have looked at the loading applied by nonliquefied crusts onto the pile cap and piles due to lateral spreading.

In such situations, Dobry *et al.* (2003) propose that in pile design, the lateral load from the nonliquefied crust plays the most important role and the contribution of the resistance offered by liquefied soil can be ignored. This will be considered in more detail in Sec. 2.8.2.

Often the nonliquefied layer that may suffer lateral spreading may be a cohesive layer. If that is the case, the lateral load applied by a cohesive crust can be determined by using shallow foundation bearing capacity factors. For example, if the clay crust overlying the liquefied layer had an undrained shear strength of S_u then the lateral pressure q applied on a rough, square pile cap in this region can simply be calculated using Eq. 2.17, following Randolph and Houlsby (1984).

$$q = (3\pi + 2) \cdot S_u \qquad (2.17)$$

For other soil types, the above expression can be suitably modified.

The lateral deflection of the pile cap and rotation of the pile cap and pile heads can be determined under the action of the lateral load induced by q above over the resisting surface area (sides of the pile cap and the portion of the piles in this region). The above expression is based on *Upper Bound Theorem of Plasticity* and, therefore, should provide a safe bound.

Alternatively, the upper bound theory by Murff and Hamilton (1993) for lateral resistance P_u can also be used in cohesive soil. In Fig. 2.5 a graph is plotted between the normalised lateral resistance $P_u/(S_uD)$ and the normalised depth z/D and D is the diameter of the pile. This method allows for a more gradual change of the undrained shear strength with depth from the surface of the soil crust to deeper regions and therefore can be used if the nonliquefied crust is reasonably deep.

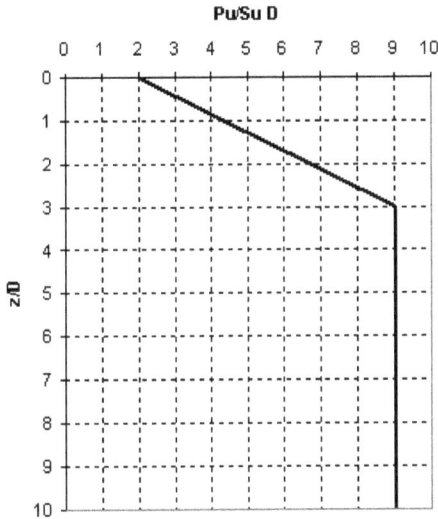

Pu/Su D

Fig. 2.5 Variation in normalised lateral load with normalised depth.

It must be noted that the lateral loading due to inertia from the superstructure and the kinematic loading due to the lateral spreading of the soil will not generally occur at the same time. However, for design purposes these can be superimposed to give a conservative assumption. The superimposed load can be used to estimate the lateral deflection of the pile heads and their rotation.

2.6 Inertial Response

The inertial response analysis uses the dynamic response obtained from the kinematic interaction study to assess the seismic displacements and rotations of the pile head or of the structure. The forces driving the pile head are derived from the mass and stiffness of the structure.

Typically the structure may be simplified as a single degree of freedom system while the piled foundation is considered to have translational and rotational degrees of freedom.

2.6.1 Relative stiffness of pile-soil system

The response of the foundation to the horizontal inertial loading and moments is determined by a combination of stiffness and damping in a manner analogous to the response of a shallow foundation. While the single pile stiffness is not sensitive to frequency, the pile group interaction terms and the radiation damping are frequency dependent.

A common way of addressing the response of the single pile or a pile group to inertial loading is by use of the concept of impedance.

$$S(\omega) = \frac{R(t)}{U(t)} \qquad (2.18)$$

where $S(\omega)$ is the (complex) impedance for the mode of response (sliding, rocking etc.) being considered, $R(t)$ is the dynamic force or moment and $U(t)$ is the dynamic displacement or rotation, respectively. $S(\omega)$ depends on the stiffness K and damping C as shown in Eq. 2.19.

$$S(\omega) = K(\omega) + i\omega C \qquad (2.19)$$

where $K(\omega)$ is the dynamic pile stiffness (kN/m), ω is angular frequency (rad/s), C is the damping coefficient (kNs/m) and $i = \sqrt{(-1)}$.

The impedance function is conveniently expressed as a complex variable because the damping component, being a function of velocity, is out of phase with the elastic stiffness. The damping may also be expressed as dimensionless frequency dependent coefficients $\zeta(\omega)$, for the various modes of response given by Eq. 2.20.

$$\zeta(\omega) = \frac{\pi fC}{K} = \frac{\omega C}{2K} \qquad (2.20)$$

This enables an alternative expression for the impedance to be developed as shown in Eq. 2.21.

$$S(\omega) = K\left[k(\omega) + 2\zeta(\omega)i\right] \qquad (2.21)$$

With impedance functions defined in the above equations, any appropriate static expression for single pile or pile group loading response can be used for the dynamic loading case, substituting the complex impedance terms for their static counterparts.

Numerical studies undertaken by Gazetas (1984) show that k(ω) is approximately unity for most practical values of pile – soil stiffness ratio over the frequencies of interest and for the horizontal, rocking and vertical modes. Hence the dynamic stiffnesses for the various modes can be taken as similar to their static counterparts.

2.6.2 Damping coefficients

Values for the damping coefficients ζ are given by Gazetas (1991) for single piles embedded in elastic soils with the stiffness profiles shown in Fig. 2.4, as shown in Table 2.3. Note that ζ_{HH} is the damping due to horizontal movement under horizontal loading, ζ_{HM} refers to horizontal movement due to applied moment and ζ_{MM} refers to rotation due to applied moment.

Table 2.3 Dimensionless pile head damping coefficients

Damping coefficient	Soil Stiffness Profile		
	Constant	Parabolic	Linear
ζ_{HH}	$0.8\beta + \dfrac{1.10\,fD}{v_s}\left(\dfrac{E_p}{E_{sD}}\right)^{0.17}$	$0.7\beta + \dfrac{1.20\,fD}{v_s}\left(\dfrac{E_p}{E_{sD}}\right)^{0.08}$	$0.6\beta + \dfrac{1.8\,fD}{v_s}$
ζ_{HM}	$0.8\beta + \dfrac{0.85\,fD}{v_s}\left(\dfrac{E_p}{E_{sD}}\right)^{0.18}$	$0.6\beta + \dfrac{0.70\,fD}{v_s}\left(\dfrac{E_p}{E_{sD}}\right)^{0.05}$	$0.3\beta + \dfrac{1.0\,fD}{v_s}$
ζ_{MM}	$0.35\beta + \dfrac{0.35\,fD}{v_s}\left(\dfrac{E_p}{E_{sD}}\right)^{0.2}$	$0.22\beta + \dfrac{0.35\,fD}{v_s}\left(\dfrac{E_p}{E_{sD}}\right)^{0.1}$	$0.2\beta + \dfrac{0.4\,fD}{v_s}$

All of the expressions in Table 3 apply only when the frequency of shaking f is greater than the natural frequency of the stratum obtained from Eq. 2.1, i.e. $f > f_n$. If the exciting frequency is below the natural frequency of the stratum then there will be no radiation damping and the damping coefficients will be the left-hand term in each case.

Calculations based on the formulae in Table 2.3 have been compared with a limited amount of field data mainly derived from experiments where vibrators have been mounted on single piles (Pender, 1993). The field data suggest that the damping coefficient values obtained from these expressions under-predict actual damping by about 30%.

Using the impedance terms, the pile head behaviour may be reduced to translational and rotational springs. The inertial loading may be determined using an idealised, single degree of freedom, structural model. The equations required to solve the response of such a system are given by Wolf (1985). Useful worked examples are given by Pender (1993).

Because the impedance terms are complex numbers, the calculated displacements also have real and imaginary parts. The maximum (real) response is readily determined by applying the SRSS (square-root of the sum of the squares) technique.

Calculations on the response of pile groups require the use of dynamic pile group interaction factors representing the effects of pile-to-pile spacing. These are frequency-dependent complex functions. Interaction factors for various loading directions and responses are given by Gazetas (1990), Gazetas *et al.* (1991) and Makris and Gazetas (1992).

2.7 *p-y* Analysis of Piles

2.7.1 *Static lateral loading*

The response of piles to static lateral loading has evolved based on the classical beams in Winkler springs theory. In the original methodology the pile-soil stiffness is modelled using nonlinear *p-y* springs derived from field tests. (Matlock and Reese, 1960; McClelland and Focht, 1958).

Let us consider a pile in a homogenous soil layer subjected to a lateral load at the pile head as shown in Fig. 2.6a. The horizontal action can be due to wind load acting on a pile or from mooring for an offshore pile founded in the sea bed. The behaviour of such a pile and the surrounding soil can be idealised by horizontal springs that will model the horizontal stiffness of the soil as shown in Fig. 2.6b. The main advantage of this method is that the nonlinear behaviour of the soil can be accommodated by means of nonlinear variation of the soil resistance p and the lateral deflection of the pile y as shown in Fig. 2.6c.

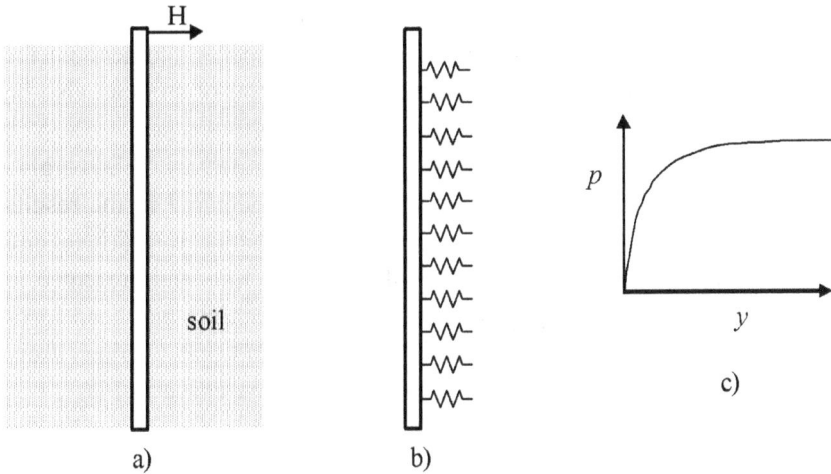

Fig. 2.6 Modelling of laterally loaded piles

The above methodology can be easily extended to layered soils. The only variation will be that the p-y curves used will vary for the different soil layers. The response of the pile to lateral loading can be obtained by using the finite differences technique to solve Eq. 2.22.

$$EI\frac{d^4y}{dx^4} = h_x \qquad (2.22)$$

where EI is the flexural rigidity of the pile, y is the lateral displacement of the pile at x, x is the distance along the pile measured from the tip and h_x is the horizontal soil force per unit length at x, which will vary as a function of y.

A similar concept called the t-z method is used to include the axial response of a pile subjected to vertical load V as shown in Fig. 2.7a. A schematic representation of a t-z model is shown in Fig. 2.7b. The axial stiffness of the pile is modelled by the springs between the pile elements. The stiffness of the soil is included by a separate set of springs on the side. The nonlinearity of the soil layers can be incorporated via a t-z curve as shown in Fig. 2.7c. As with p-y analysis, soil layers with different t-z curves can be incorporated into the analysis to account for layered soil deposits. Similarly, strength degradation of soil can also be incorporated.

The applicability of t-z analysis was demonstrated by Vijayvergiya (1977) who proposed a t-z curve that could be compared to the results from a pile load test on a 400mm diameter steel tube pile driven into medium dense sand. This result is re-plotted in SI units and presented in Fig. 2.8 which shows very good correlation between the pile head load-settlement curve obtained from the t-z analysis and that obtained from the pile load test. Vijayvergiya proposed that the t-z curve in Fig. 2.8 is suitable for piles of up to 0.6m diameter. For larger diameter piles the soil springs need to be modified to prevent an over-stiff response.

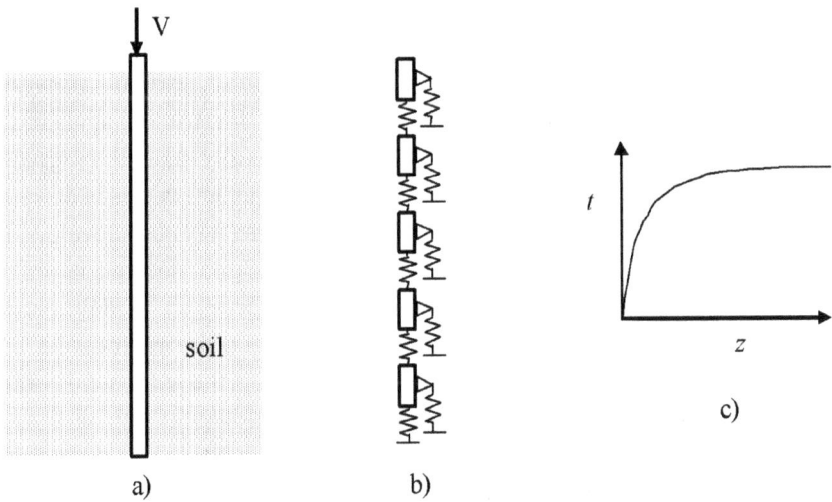

Fig. 2.7 Modelling of axially loaded piles.

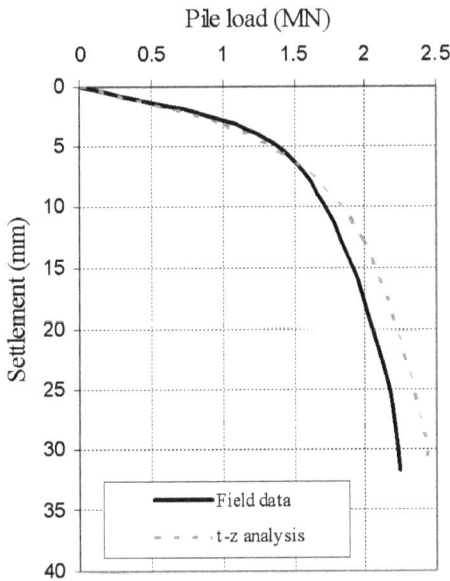

Fig. 2.8 Comparison of *t-z* analysis to actual pile load tests in sand (re-plotted in SI units from Vijayvergiya, 1977).

2.7.2 Cyclic lateral loading

The use of the '*p-y* method' has become very popular, particularly with the oil industry where the main requirement is for use in analysing laterally loaded offshore piles. These piles generally extend to great depths below the seabed and are subjected to extreme lateral loads during storm conditions. The American Petroleum Institute (API) has developed *p-y* curves that can be used in the design of such piles located in soft, clayey soils, API Code RP2A (1984). The API curves are widely used for short term static loading and cyclic loading and are reproduced in Fig. 2.9. In this figure the axes are normalised by the ultimate lateral resistance offered by the soil p_u and the limiting horizontal displacement y_u. It can be seen from this figure that the soil resistance is degraded severely with increasing lateral deflection of the pile in the presence of cyclic loading. This is in contrast to the static loading case where full lateral resistance is mobilised at large lateral deflections.

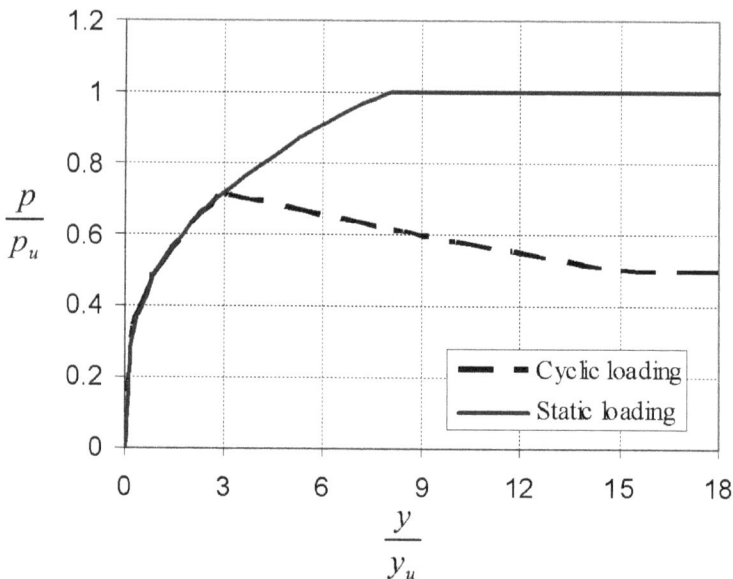

Fig. 2.9 API curves for soft clay

For the purposes of Fig. 2.9, the ultimate lateral deflection y_u can be estimated from the data from an undrained triaxial compression test and using Eq. 2.23.

$$y_u = 2.5\varepsilon_{50}D \qquad (2.23)$$

where D is the diameter/width of the pile and ε_{50} is the strain at 50% of the maximum stress in an undrained triaxial compression test.

2.7.3 p-y analysis under earthquake loading – level ground

The *p-y* methodology developed for laterally loaded piles has been extended to the analysis of piles subjected to earthquake loading. During an earthquake the soil layer will suffer displacements. A pile passing through the soil layer will also suffer displacements which are different from the free-field displacements as demonstrated in Fig.2 earlier. In contrast to the case of static lateral loads applied to piles discussed in Sec. 2.7.1, the *p-y* analysis for the earthquake loading case would consider relative horizontal displacement of the pile with respect to the soil at any given point along the pile. So the soil resistance *p* at any location is related to the relative displacement of the pile *y*. Once this distinction is made the same methodology as discussed in Sec. 2.7.1 can be used.

Two aspects are to be noted in the context of *p-y* analysis applied to piles in level ground subjected to earthquake loading. Firstly, the bending deformation mode of the pile can become very important as this will influence the lateral displacements of the pile relative to the soil. For example, as illustrated in Figs. 2.1 and 2.2, the inertial and kinematic loads can cause different bending mode shapes depending on the following factors:

i.   relative stiffness of pile with respect to the soil;
ii. translation and/or moment restraint condition at the pile top i.e. whether the pile cap is restrained from translation and/or rotation;
iii. moment fixity offered at the pile tip;
iv. fixity length required when a pile passes from a loose soil layer to a stiff soil layer.

Accordingly, the relative displacement of the pile with respect to the soil can change thereby affecting the magnitude and location of the soil resistance p offered by the soil.

The second aspect of p-y analysis that needs to be considered carefully is if one or more soil layers through which the pile is passing suffers liquefaction during an earthquake event. In such cases it is important to consider the degradation in the soil stiffness and strength owing to accumulation of shear strain and/or generation of excess pore pressure. Similarly any changes in the excess pore pressure due to dilation of soil, say, when the stress path crosses the Phase Transformation Line should be considered. (Ishihara, 1993). For example, a strong cycle of shaking of an already liquefied ground can cause local dilation of soil which is manifested as suction (drop in excess pore pressure). This will cause a temporary stiffening of the soil which, at that instant, can offer an increased soil resistance p to the pile. This is particularly important if the transient loading on the pile exceeds the residual loading at the end of the earthquake event. These aspects are considered in further detail in Chapters 3 and 4.

2.7.4 *p-y analysis under earthquake loading – sloping ground*

Sloping ground often presents the additional problem of lateral spreading particularly if there is a nonliquefied crust on the surface, as discussed above. Before considering the nonliquefied crust, let us consider the case of a sloping ground that suffers liquefaction and hence undergoes lateral spreading. The p-y methodology presented in Sec. 2.7.3 can be extended to include the effects of lateral spreading in sloping ground. Goh and O'Rourke (1999) developed a set of normalised p-y curves for smooth and rough piles as re-plotted in Fig. 2.10. These curves were developed based on a series of finite element analyses and the authors show good agreement of these curves with dynamic centrifuge test results conducted by Abdoun (1997) at RPI. In Fig. 2.10 the soil resistance p is normalised by the undrained shear strength S_u and the pile diameter D. The relative pile–soil displacement y is also normalised by the pile diameter D. In this figure the theoretical upper bounds for the smooth and rough circular piles proposed by Randolph and Houlsby (1984) are also shown. Figure 2.10 illustrates the strain softening behaviour of the laterally spreading soil for both smooth and rough piles.

While designing piles in layered soils on a slope, the additional loading from nonliquefied crusts must be considered. The nonliquefied layers can be treated as additional layers with an appropriate *p-y* response. If the nonliquefied layer consists of stiff, over-consolidated clay then the *p-y* curve may include quite a stiff response for small lateral displacements followed by a softer response at large displacements, similar to the *p-y* curves in Fig. 2.10. On the other hand, if the nonliquefied layer is soft clay, then the *p-y* curve may include a soft response. These are shown schematically in Fig. 2.11.

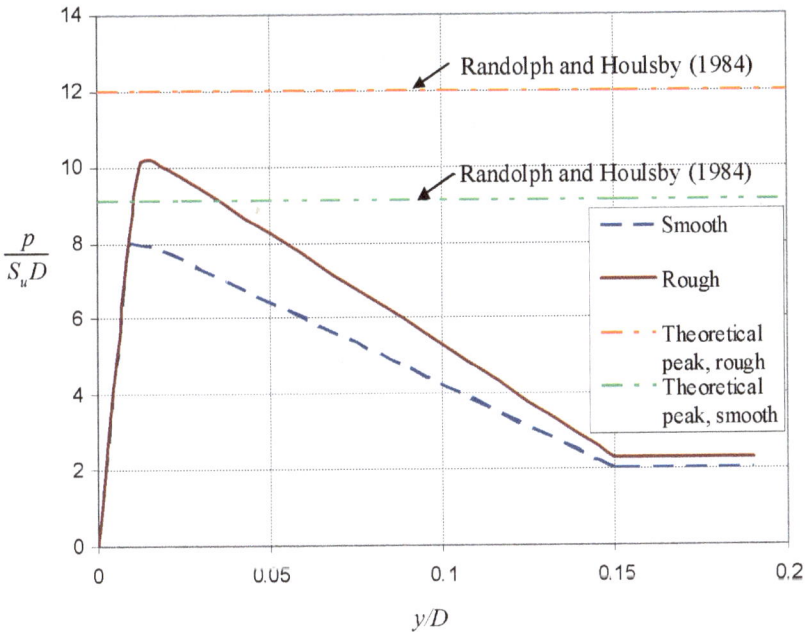

Fig. 2.10 *p-y* curves for laterally spreading soil, re-plotted from Goh and O'Rourke (1999).

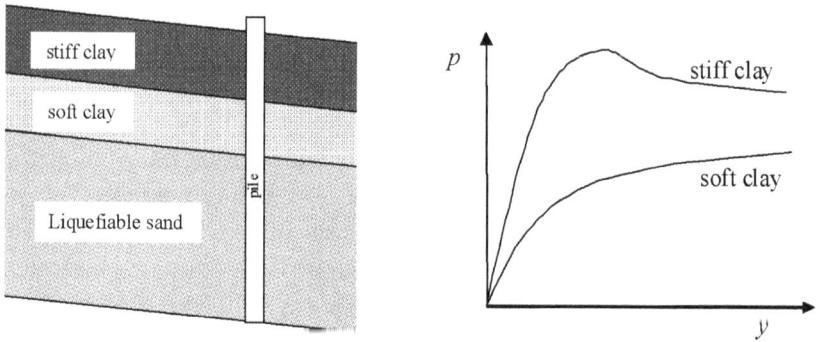

Fig. 2.11 *p-y* curves for stiff and soft clay.

As in the case of level ground, the transient loading on the piles needs to be considered while designing the piles to ensure their safety during the earthquake loading itself. So the actions on the pile that arise due to the dilation of liquefied ground and its subsequent stiff response must be included in the *p-y* response. Haigh (2002) proposed inclusion of excess pore pressure as a variable to modify the *p-y* curves for liquefiable soil layers. A family of *p-y* curves is considered, in which higher curves are associated with weak events that result in low excess pore pressure ratio r_u (low strength earthquakes), while lower curves are used in the case of strong events that result in high excess pore pressure ratio r_u (medium to strong earthquakes). This modification of the traditional *p-y* analysis is very useful to allow for soil strength degradation with excess pore pressure generation.

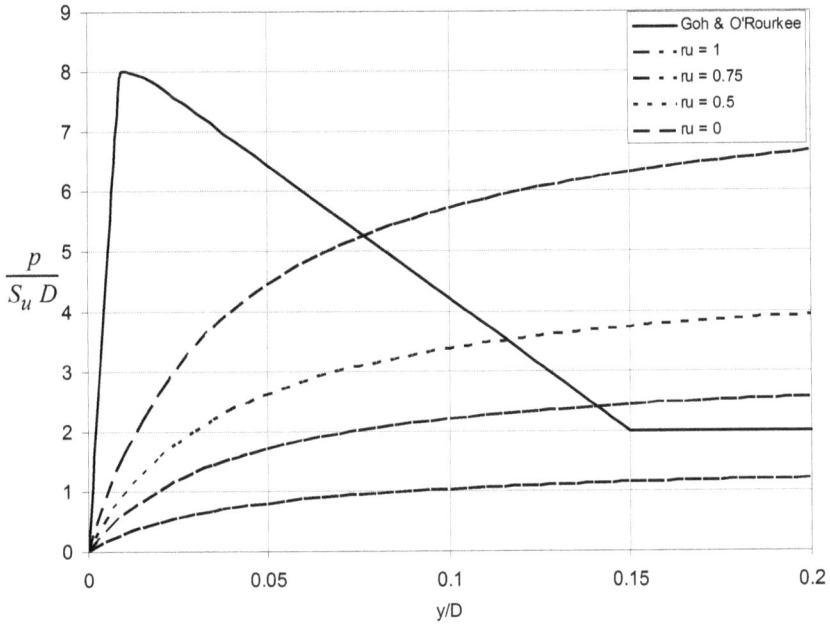

Fig. 2.12 Modified *p-y* curves for liquefiable soils.

2.8 Limit Equilibrium Analysis of Piles Subjected to Earthquake Loading

Limit equilibrium analyses are quite popular in the design of retaining walls, shallow foundations and other geotechnical structures. It is possible to carry out limit equilibrium analyses to calculate the bending moments and shear forces in piles in laterally spreading soils. Such analyses are considered as an alternative to the *p-y* analysis discussed in Sec. 2.7 and would generally involve further assumptions.

2.8.1 Limit equilibrium of piles in laterally spreading soils

Let us first consider the simple case of single piles in laterally spreading soils overlying a dense soil layer, as shown in Fig. 2.13a. There is no nonliquefied crust above the spreading soil. In this case it is straightforward to establish a limit equilibrium model using the free body diagram shown in Fig. 2.13b. This type of analysis was presented by Dobry *et al.* (2003) for relatively stiff piles. The results from the limit equilibrium analysis were validated against centrifuge tests conducted in loose deposits of saturated Nevada sand with the dense layer simulated by lightly cemented Nevada sand. (Abdoun *et al.*, 2003).

Let the soil resistance per unit area offered by liquefied soil be p_l. Various researchers have estimated the value of p_l to be between 8kPa and 20kPa based on centrifuge tests on flexible and rigid piles in laterally spreading soils and based on centrifuge tests in which shear wave velocity measurements were made in liquefied soils. (Dobry *et al.*, 2003; Haigh, 2002; Ghosh and Madabhushi, 2002).

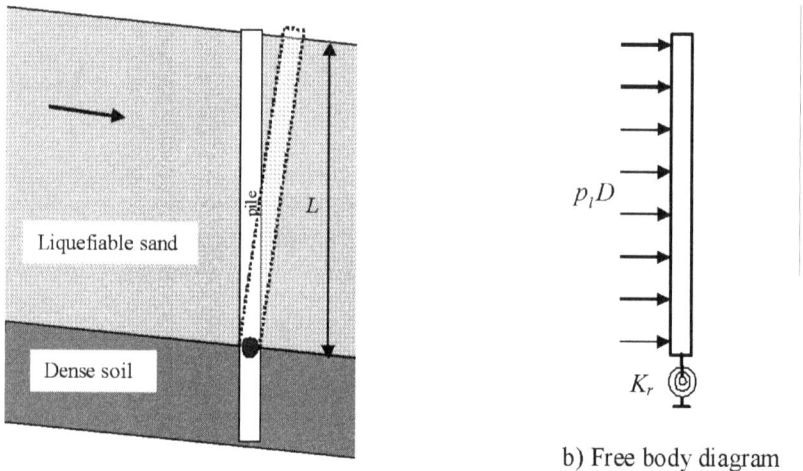

a) single pile configuration

b) Free body diagram

Fig. 2.13 Single pile in laterally spreading soil.

Using this, the force applied by the liquefied soil on the pile can be calculated following Eq. 2.24.

$$F_p = p_l DL \qquad (2.24)$$

The maximum bending moment will clearly occur at the interface with the dense soil and its magnitude will be;

$$M_{max} = \frac{1}{2} p_l DL^2 \qquad (2.25)$$

If a pile cap of resisting area A_c is present at the top of the pile within the laterally spreading soil, the force acting on the pile will increase by $p_l A_c$. The maximum bending will also increase as shown in Eq. 2.26.

$$M_{max} = p_l L \left[\frac{1}{2} DL + A_c \right] \qquad (2.26)$$

The rotational spring K_r is not required for estimating the maximum bending moment. However, a value of K_r is required if the lateral displacement of the pile top is required. Dobry *et al.* (2003) quote a value of 5738kNm/rad as the rotational stiffness for lightly cemented Nevada sand used in their centrifuge tests.

As an example, if we consider a 0.5m diameter passing through a 10m thick liquefied layer suffering lateral spreading and entering into lightly cemented dense sand below, the maximum bending moment can be calculated using Eq. 2.26. Let us assume that the upper bound value of p_l = 20kPa is exerted by the liquefied soil on the pile.

$$M_{max} = \frac{1}{2} \times 20 \times 0.5 \times 10^2 = 500 kNm \qquad (2.27)$$

If the lightly cemented dense sand is assumed to have a rotational stiffness of 5738kNm/rad quoted by Dobry *et al.* (2003), then the lateral displacement at the pile top can be calculated as

$$\delta_{top} = \frac{M_{max}}{K_r} L = \frac{500}{5738} \times 10 = 0.87m \qquad (2.28)$$

Clearly this is quite a significant lateral displacement for the pile. In addition it should be noted that such large lateral displacements can be amplified significantly by any superstructure load acting on the pile giving rise to large $P - \Delta$ effects on the pile. This aspect was discussed by Brandenberg *et al.* (2005b) following research conducted at Cambridge on pile instability by Bhattacharya *et al.* (2004, 2005) and Knappett and Madabhushi (2005). Generally, piles by their very nature will be carrying large axial loads from the superstructure and therefore it is very important to consider the effects of axial loading on the piles on the stability of the pile as the soil suffers liquefaction and undergoes lateral spreading. These aspects are considered in further detail in Chapters 3 and 5.

2.8.2 Limit equilibrium analysis in the presence of nonliquefied crust

2.8.2.1 Stiff clay as a nonliquefiable layer

Let us now consider the case of a single pile of diameter D passing through a stiff clay layer of thickness h and a liquefiable sand layer of thickness L before entering dense, nonliquefiable soil as shown in Fig. 2.14. As described in Sec. 2.1.3 and seen in the case histories in Sec. 1.5 of Chapter 1, this type of soil stratification is susceptible to lateral spreading and the stiff clay layer above the liquefiable sand will undergo lateral spreading and is capable of subjecting the pile to large lateral loads as passive soil pressures are generated.

Dobry *et al.* (2003) describe a limit equilibrium-based procedure for a similar soil configuration for a relatively flexible pile passing through a lightly cemented soil layer at the top and then through a thicker layer of liquefiable sand before entering yet another layer of nonliquefiable material at the bottom, also made up of lightly cemented sand. The same procedure can be modified for the case of a stiff clay layer at the top acting as the nonliquefiable crust. The general procedure of limit equilibrium analysis is described first before discussing some of the limitations.

When the sloping ground suffers lateral spreading following an earthquake event, both the liquefiable layer and the stiff clay layer on the top will suffer lateral spreading. This will cause the pile to deform as shown schematically in Fig. 2.14a in dashed lines. As in Sec. 2.8.1, free body diagrams of the soil pressures and hence forces acting on the pile can be sketched as shown in Fig. 2.14b. For this analysis let us consider points A and B on the pile at the top and bottom of the liquefiable layer. Let point D be the top of the pile. At a certain depth z_{ps} below the top of the pile, a point C is considered. This point C demarcates a shallow region of the stiff clay layer which applies a passive pressure from the down slope side on to the pile and the deeper region of the stiff clay layer which applies a passive pressure from the upslope side on to the pile as indicated in Fig. 2.13b. Further, a rotational spring with stiffness K_r is assumed at point B similar to the one considered in Sec. 2.8.1. This spring models the resistance offered by the lower layer of nonliquefiable soil to the pile.

Let us assume that the stiff clay layer above the liquefiable layer has a unit weight of γ and undrained shear strength of S_u. Consider the passive soil pressures in this layer at points D, C and A. From point D to C the passive pressure varies linearly from $2S_u$ to $2S_u + \gamma z$, and acts from right to left. Similarly from points C to A the passive pressure varies from $2S_u + \gamma z$ to $2S_u + \gamma h$ and acts from left to right on the pile.

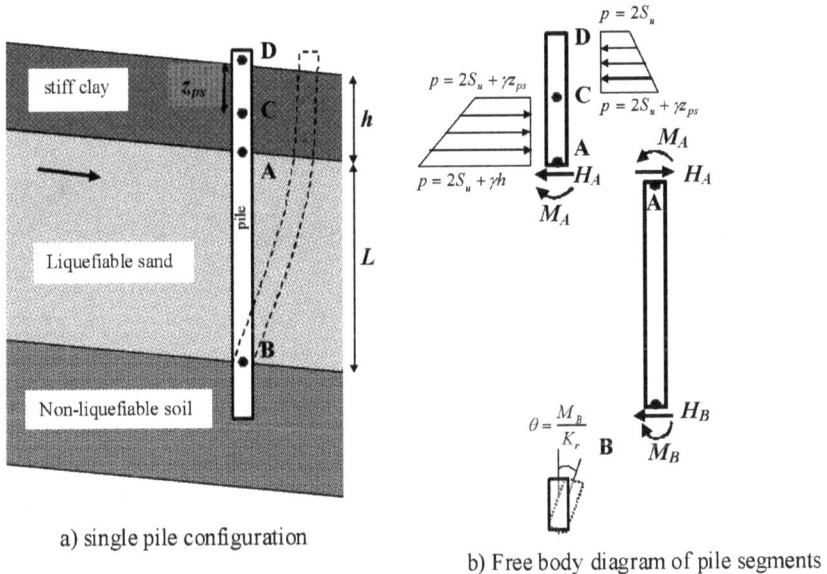

a) single pile configuration

b) Free body diagram of pile segments

Fig. 2.14 Single pile passing through a laterally spreading soil and a nonliquefied crust.

Dobry *et al.* (2003) assume that for flexible piles when the nonliquefied layer on the top is stiff, the following assumptions can be made.

i.   The pile remains in the elastic range and no plastic hinging occurs in the piles.
ii. Only small to moderate amounts of lateral spreading occur.
iii. The liquefiable layer itself exerts very little lateral pressure on the pile during lateral spreading, such that
iv. $M_A \backsim M_B$ i.e. moment loading acting on the top and bottom of the pile segment passing through the liquefiable layer are approximately equal.

With these assumptions, the following equations can be written from limit equilibrium considerations.

$$H_A = \left[2S_u + \frac{\gamma}{2}(h + z_{ps})\right](h - z_{ps})D - \left[2S_u + \frac{\gamma}{2}z_{ps}\right]z_{ps}D \qquad (2.29)$$

$$H_B = H_A \qquad (2.30)$$

$$M_A = 2DS_u z_{ps}\left[h - \frac{z_{ps}}{2}\right] + \frac{D}{2}\gamma z_{ps}^2\left[h - \frac{2}{3}z_{ps}\right]$$

$$-\frac{D}{2}\left[2S_u + \gamma z_{ps}\right]\left[h - z_{ps}\right]^2 - \frac{D}{6}\gamma(h - z_{ps})^3 \qquad (2.31)$$

$$M_B = M_A - H_A L \qquad (2.32)$$

The four static equilibrium equations given as Eqs. 2.29–2.32 are not sufficient to calculate the five unknowns H_A, H_B, M_A, M_B and z_{ps}. Dobry *et al.* (2003) suggest that the bending moments in the pile at the top and bottom of the liquefiable layer are approximately equal, $|M_A| \approx |M_B|$, and use this as the extra condition for solving the equations 2.29 to 2.32. This assumption was based on centrifuge test data reported by Abdoun *et al.* (2003) and is perhaps valid only for flexible piles and relatively stiff nonliquefiable layer as pointed out by the discussion of Brandenberg *et al.* (2005b).

Limit equilibrium analyses are generally carried out using the 'worst case loading' scenario. It must be pointed out that in the above analysis the worst case loading occurs when $z_{ps} = 0$. Brandenberg *et al.* (2005b) comment on this worst case 'Limit State' and suggested modifications to the above equilibrium equations, for the case of lightly cemented soil forming the nonliquefiable crust reported by Abdoun *et al.* (2003). Similar modifications are shown below for the case considered here based on a stiff clay layer forming the nonliquefiable crust.

When z_{ps} = 0, there are no passive earth pressures acting on the pile from right to left as shown in Fig. 2.15a and b. The pressure exerted by the liquefied soil layer p_l can also be easily incorporated into the analysis as shown in Fig. 2.15b. The pile may suffer deflections as suggested by the dashed line in Fig.14a. This will of course depend on the stiffness of the nonliquefiable layer, the flexural stiffness of the pile and the geometry of the problem, i.e. the thicknesses of the nonliquefied crust h and the thickness of the liquefiable layer L. For the limit state shown in Fig. 2.15 the following equilibrium equations can be written.

$$H_A = hD\left[2S_u + \frac{1}{2}\gamma h\right] \tag{2.33}$$

$$H_B = hD\left[2S_u + \frac{1}{2}\gamma h\right] + p_l L D \tag{2.34}$$

$$M_A = hD\left[S_u h + \frac{1}{6}\gamma h^2\right] \tag{2.35}$$

$$M_B = hD\left[S_u h + \frac{1}{6}\gamma h^2\right] + hDL\left[2S_u + \frac{1}{2}\gamma h\right] + \frac{1}{2}p_l DL^2 \tag{2.36}$$

For a given geometry of the problem, i.e. h, L and D and knowing the undrained shear strength of the stiff layer S_u and estimating the pressure exerted by the liquefied sand layer p_l using the range of values observed in the centrifuge tests (8kPa to 20kPa) as discussed in Sec. 2.8.1, the above four equations can be used to calculate H_A, H_B, M_A and M_B. The problem has now become determinate as there is no fifth unknown z_{ps}.

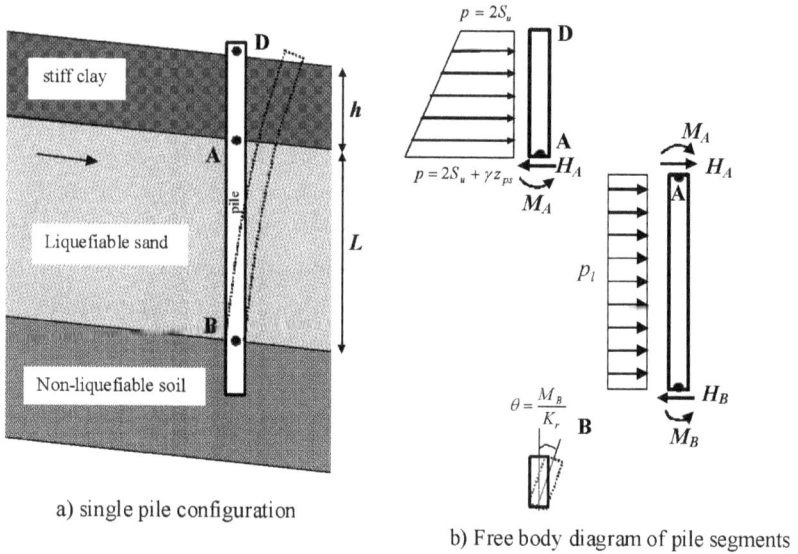

a) single pile configuration

b) Free body diagram of pile segments

Fig. 2.15 Limit State for a single pile passing through laterally spreading stiff clay layer.

2.8.2.2 Dense sand as a nonliquefiable layer

Another common condition that is encountered in the field is a dense sand layer above the liquefiable layer. This can occur in the field due to natural deposition or densification efforts prior to construction. The water table could be present at a depth of h, thereby rendering the shallow soil layer above this to be nonliquefiable. A typical configuration is shown in Fig. 2.15. For this case let us include a fixing moment M_{Top} at the pile head. This moment could arise due to resistance to rotation offered by the framing action of a pile group and/or the rotational restraint offered by the superstructure.

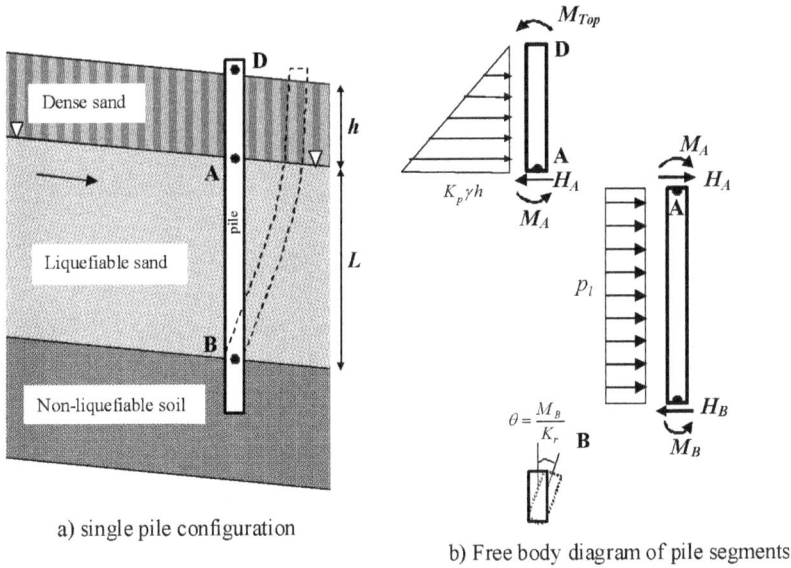

a) single pile configuration

b) Free body diagram of pile segments

Fig. 2.16 Limit State for a single pile passing through laterally spreading dense sand layer.

For the limit state shown in Fig. 2.16 the following equilibrium equations can be written, assuming that full passive earth pressure coefficient K_p in the dense sand layer is mobilised at limit state.

$$H_A = \frac{1}{2} K_p \gamma h^2 D \qquad (2.37)$$

$$H_B = \frac{1}{2} K_p \gamma h^2 D + p_l L D \qquad (2.38)$$

$$M_A = \frac{1}{6} K_p D \gamma h^3 - M_{Top} \qquad (2.39)$$

$$M_B = \frac{1}{6} K_p D \gamma h^3 + \frac{1}{2} K_p \gamma h^2 L D + \frac{1}{2} p_l D L^2 - M_{Top} \qquad (2.40)$$

Let us consider a practical example for this case. Let us assume a 3m thick nonliquefiable layer with a unit weight of 16kN/m^3 and peak friction angle of 35°. This layer overlies a 10m thick liquefiable sand layer. Consider a pile which has 0.5m diameter passing through these layers and entering into a nonliquefiable dense sand layer below. Assume the upper bound value for the pressure applied by the liquefied sand p_l of 20kPa. Let us also assume a resisting moment of 200kNm from the pile cap applied at the pile head.

$$K_p = \frac{1+\sin\phi}{1-\sin\phi} = 3.7 \tag{2.41}$$

Therefore, using Eqs.2.37 and 2.38 we can calculate H_A and H_B as

$$H_A = \frac{1}{2} \times 3.7 \times 16 \times 3^2 \times 0.5 = 133.2kN \tag{2.42}$$

$$H_B = H_A + p_l DL = 133.2 + 20 \times 0.5 \times 10 = 233.2kN \tag{2.43}$$

Similarly, using Eqs. 2.39 and 2.40 and the value of resisting moment from pile cap at the pile head, we can calculate M_A and M_B as

$$M_A = \frac{1}{6} \times 3.37 \times 0.5 \times 16 \times 3^3 - 200 = -78.7kNm \tag{2.44}$$

$$M_B = -78.7 + 133.2 \times 10 + \frac{1}{2} \times 20 \times 0.5 \times 10^2 - 200 = 1553.3kNm \tag{2.45}$$

Thus it is straightforward to use the limit equilibrium equations to determine the shear force and bending moment distributions in pile foundations.

2.9 Provisions in Eurocode 8

Eurocode 8 – Part 5 (2003), which deals with geotechnical aspects of earthquake engineering, has certain provisions applicable to the design of pile foundations in seismic regions. This code also has a methodology to determine the liquefaction potential of a site using a modified form of the Seed and Idriss type chart between the cyclic resistance ration (CRR) and the corrected SPT N values. The general pile design under static loading both axial and lateral loading is covered by Eurocode 7 (1997) on geotechnical design. It must be pointed that Eurocode 8 does not consider exclusively the design of piles passing through liquefiable soil deposits. Some of the useful guidelines offered by Eurocode 8 are presented here.

2.9.1 Combination rules

Since the response of the soil and structure will be at different natural frequencies, the combination rules given in Clause 4.3.3.5 of Eurocode 8 – Part 1 (2003) that deals with seismic actions can be used to calculate the cumulative effect of kinematic and inertial interaction.

2.9.2 Pile head fixity coefficients

Eurocode 8 – Part 5, Annex C provides guidance on the pile head stiffness coefficients for the three types of idealised soil stiffness profiles presented earlier in Fig. 2.4. These are reproduced in Table 2.3 and can be used in preference to the flexibility coefficients presented earlier in Sec. 2.2.2, Table 1. As before, the key parameters are: E is the Young's modulus of the soil (= $3G$, undrained loading assumed), E_p is the Young's modulus of the pile material, E_s is the Young's modulus of the soil at one pile diameter depth, D is the pile diameter and z is the pile depth.

Table 2.3 Static stiffness of flexible piles embedded in three soil models.

Soil Model ↓	K_{HH}/DE_s	K_{MM}/D^3E_s	K_{HM}/D^2E_s
Linear variation $E = E_s \dfrac{z}{D}$	$0.60\left(\dfrac{E_P}{E_s}\right)^{0.35}$	$0.14\left(\dfrac{E_P}{E_s}\right)^{0.80}$	$-0.17\left(\dfrac{E_P}{E_s}\right)^{0.60}$
Square Root variation $E = E_s \sqrt{\dfrac{z}{D}}$	$0.79\left(\dfrac{E_P}{E_s}\right)^{0.28}$	$0.15\left(\dfrac{E_P}{E_s}\right)^{0.77}$	$-0.24\left(\dfrac{E_P}{E_s}\right)^{0.53}$
Constant $E = E_s$	$1.08\left(\dfrac{E_P}{E_s}\right)^{0.21}$	$0.16\left(\dfrac{E_P}{E_s}\right)^{0.75}$	$-0.22\left(\dfrac{E_P}{E_s}\right)^{0.50}$

Once the loads on the pile heads and the pile cap are determined as discussed in Sec. 2.3, the stiffness coefficients in Table 2.3 can be used to determine the lateral and cross pile head displacements and the pile cap rotation.

A few points need to be made here to clarify the above flexibility coefficients. Eurocode 8 still uses the Young's modulus of soil E_s as seen in Table 2.3. As discussed earlier this parameter is not a satisfactory parameter for soil stiffness, which is a highly nonlinear parameter. Further, the Young's modulus of soil E_s for the purposes of Table 2.3, is to be estimated at a depth of one pile diameter. The mean effective confining stress at such shallow depths will be very small and therefore the definition and use of E_s at such depths is doubly disappointing.

Secondly, the use of Young's modulus of the pile material E_p also warrants some forethought. As it is the pile stiffness one is after, use of Young's modulus instead of flexural rigidity EI needs to be implemented carefully in design calculations. For example, if a hollow, steel, tubular pile is being designed then the Young's modulus of the pile needs to be corrected to account for the hollow pile. This can be achieved by using the following equations, for example.

$$E_{p\_corrected} = \frac{E_p}{\left(I_{solid}/I_{tubular}\right)} \tag{2.46}$$

Where I_{solid} is the second moment of area of an equivalent solid pile and $I_{tubular}$ is the second moment of area of a tubular pile.

Eq. 2.46 can be simplified for a hollow, circular pile into Eq. 2.47 below.

$$E_{p\_corrected} = \frac{E_p}{\left(D_o^4 / \left\{D_o^4 - D_i^4\right\}\right)} \tag{2.47}$$

where D_o is the diameter of a solid pile and the outer diameter of tubular pile and D_i is the inner diameter of a tubular pile.

2.9.3 Kinematic loading

Eurocode 8 – Part 5 (2003) notes that bending moments due to kinematic interaction only need to be considered when all the following conditions apply.

i. Ground profile is of type D, S_1 or S_2 and contains consecutive layers of sharply differing stiffness.
ii. The zone is of moderate or high seismicity (i.e. αS exceeds 0.1g), and the structure is of importance class III or IV.

Further, Eurocode 8 – Part 5 requires piles to remain elastic, though under certain conditions they are allowed to develop plastic hinges at their heads. The regions of plastic hinging should be designed according to Eurocode 8 – Part 1 (2003), Clause 5.8.4.

2.10 Summary

Earthquake loading on pile foundations can be mainly classified as inertial and kinematic loading. Inertial loading on the pile arrives from the superstructure while the kinematic loading on the pile is generated by relative displacements between the pile and surrounding soil during and immediately following the earthquake. In this chapter inertial and kinematic loadings on pile foundations were considered for idealised soil profiles. Methods of analysis were presented firstly for inertial loading based on pile flexibility coefficients. Methods for estimating the kinematic loading in level ground and sloping ground were also presented.

p-y analysis is very popular for piles subjected to static lateral loading. This method is similar to *t-z* analysis for determining the axial response of piles. Both of these analysis methods are introduced in this chapter. Extension of the *p-y* analysis method for the case of earthquake loading on piles was discussed both for the simple level ground case and for sloping ground suffering lateral spreading. Use of limit equilibrium methods for estimating the maximum bending moments and shear forces in piles were introduced for the case of sloping ground both with and without a nonliquefiable crust. Finally, the provisions of Eurocode 8 that deal with the design of structures subjected to earthquake loading were introduced, particularly those relevant to pile foundations. The modifications suggested by Eurocode 8 to pile flexibility coefficients have been presented.

Chapter 3

Accounting for Axial Loading in Level Ground

3.1 Liquefaction as a Foundation Hazard

3.1.1 Liquefaction

When saturated soil is subjected to seismic shaking, densification leads to a reduction in pore volume. If this happens more rapidly than the pore water can flow out of the compressed voids, pore water pressure increases. When the excess pore pressure generated in this manner is equal to the initial effective stress, the inter-granular effective stress will be negligible and the soil is said to be liquefied. The degree of the rise in the excess pore pressure can be defined by the *excess pore pressure ratio* (r_u)

$$r_u = \frac{\Delta u}{\sigma'_{v0}} \tag{3.1}$$

where Δu is excess pore pressure and σ'_{v0} the initial effective stress. When loose sands become fully liquefied ($r_u = 1$), they can sustain large shear strain (flow) at constant volume. (Ishihara, 1993). If a structure is to be founded on liquefiable ground it is often desirable to use piles as these may be designed to extend through the liquefiable soil to bear in more competent underlying layers. In this way, the structural weight that is carried in the piles in the form of axial load may be transmitted to a more resilient stratum of soil to avoid large liquefaction-induced settlements which may be associated with shallow foundations.

As liquefiable soil is commonly sandy in composition, the more competent bearing layers underlying liquefiable soil deposits tend to also be sandy, but of increased density. Considering the mechanism by which excess pore pressure is generated at the micro-mechanical scale, dense sands are commonly assumed to remain competent during strong shaking. Coelho *et al.* (2003) and Elgamel *et al.* (2005) have shown by dynamic centrifuge testing that significant excess pore pressures can be generated due to seismic shaking in dense saturated sands. Unlike in loose sands however, applied shear strain leads to a reduction in excess pore pressure and consequent stiffening. Large permanent displacement may be accumulated from small cyclic incremental shear strains. This process is called *cyclic mobility*. (Castro, 1975; Castro and Poulos, 1977).

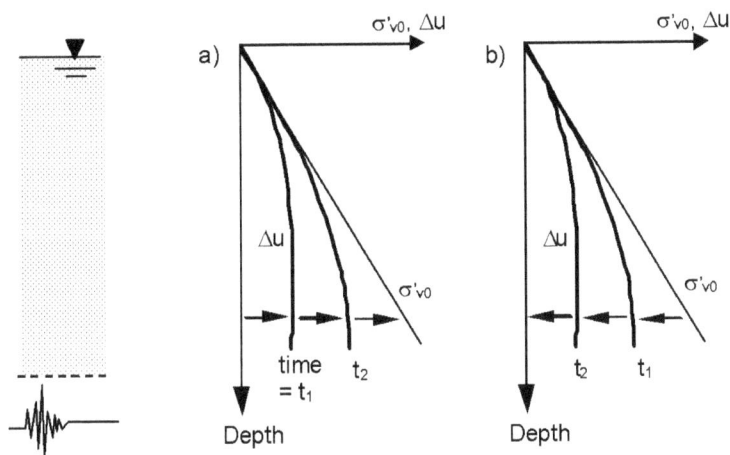

Fig. 3.1 Progression of liquefaction front within a shaken column of saturated soil (a) liquefaction, front moving from top-to-bottom (b) reconsolidation, front moving upwards.

Due to the increase of effective stress with depth in the ground, liquefaction proceeds in a 'top-down' fashion as the upper layers of soil require a smaller increase in excess pore pressure to become fully liquefied (Fig. 3.1). This was originally shown by Florin and Ivanov (1961), and leads to the concept of a liquefaction front which moves downwards in the soil as shaking progresses (Fig. 3.1a). For a soil stratum which is liquefied through its entire depth, the excess pore pressures will be higher at the bottom than at the top so that there will be an upwards hydraulic gradient within the soil. Following the end of shaking, this gradient will cause the soil to reconsolidate from the bottom up (Fig. 3.1b).

3.1.2 Determination of liquefaction susceptibility

The extent to which liquefaction is expected to occur in a given soil profile is often accomplished from SPT or CPT tests and empirical correlations. One of the most commonly used methods is that of Seed and Idriss (1971), which, in its current form can incorporate the effects of soil overburden and fines content and can be used with both SPT N-value and CPT tip resistance (q_c). (Idriss and Boulanger, 2004). These correlations plot the Cyclic Resistance Ratio (CRR) which is the cyclic stress ratio causing full-liquefaction for a magnitude 7.5 earthquake against normalised SPT N-value, $(N_1)_{60}$ or CPT tip resistance. For a given earthquake, the induced Cyclic Stress Ratio (CSR) is calculated. If CSR >CRR then the soil is considered to be fully liquefied. Due to the analogy between piles and cones, CPT data is commonly preferred in the design of piled foundations. As a result, this book will focus on the use of CPT data for liquefaction evaluation.

These empirical methods are limited, however, as they can lead to loose sands being classed as 'liquefiable' while dense sands are considered 'nonliquefiable' due to their higher cone resistance. An example of this is shown in Fig. 3.2, in which the values of CSR computed for the two depths shown correspond to a magnitude 7.5 earthquake causing a peak horizontal acceleration of 0.2g. In this example, the correlation suggests that the loose soil will be problematic as it will fully liquefy. However, it is not possible to say anything about the degree of excess pore pressure rise within the dense layer, other than that r_u will not reach a value of one. As will be seen later in this chapter, the development of significant excess pore pressures in dense soils (commonly used as pile bearing layers) can lead to bearing capacity failure and excessive settlement of piles, even if r_u does not reach a value of one in these layers.

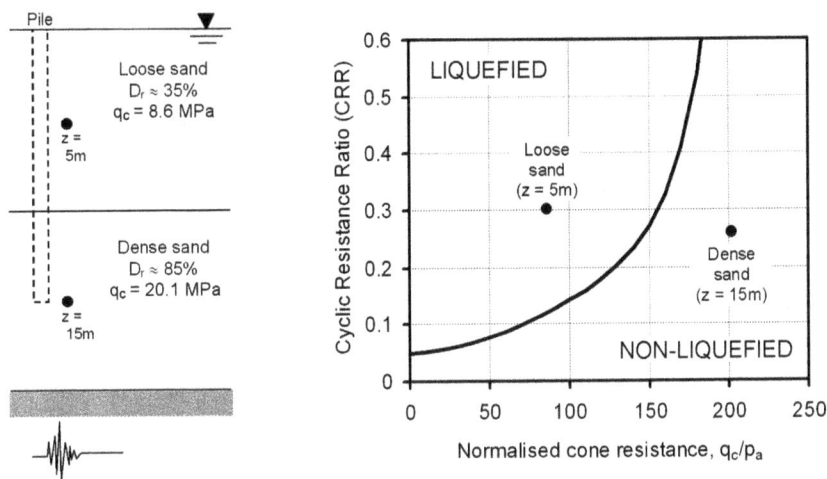

Fig. 3.2 Example showing the limitations of empirical methods for determining the hazard posed to piled foundations by liquefaction (N.B. p_a = atmospheric pressure = 100kPa).

It is now possible to undertake more sophisticated site response analyses to determine full time histories of excess pore pressure generation in heavily non-uniform soil profiles. This is made possible using advanced constitutive models for liquefaction such as that proposed by Yang *et al.* (2003). This particular model has been incorporated into a simple site response tool, CYCLIC 1D (http://cyclic.ucsd.edu/), which allows designers to determine the excess pore pressure at any depth within a body of soil at a given site and for a given earthquake. The ability to make such estimations of expected excess pore pressures is important as the following discussion of the effects of axial load on pile behaviour in liquefiable soil will link pile behaviour to the excess pore pressures developed within the surrounding ground.

3.2 Influence of Axial Loading on Pile Failure

The design of piles under static conditions essentially involves selecting pile dimensions and axial loads to ensure that there is an adequate factor of safety (FOS) against bearing failure and that excessive settlement will not occur. The layout of the foundation may then be optimised to ensure that the piles support the structure in the most efficient way possible, thereby minimising the cost of the foundation.

Under seismic conditions the effects of lateral inertial and kinematic loading on the pile (c.f. Chapter 2) are often assumed to dominate over axial loads in terms of inducing pile failure. Additional design work will then concentrate on determining the bending moments and shear forces generated within the pile due to combinations of inertial and kinematic loads, which are checked to ensure that bending or shear failure of the piles does not occur. In liquefying soil however, the changes in effective stress around a pile can lead to additional axial modes of failure when the pile carries significant axial load.

For end-bearing piles the reduction in effective stress around the pile shaft and base can lead to a reduction in shaft and base capacity and excessive settlement (Knappett, 2006). This is discussed further in Secs. 3.3 to 3.5. High excess pore pressures around the pile shaft can also lead to a reduction in lateral pile support. If the soil beneath the pile retains some of its strength (e.g. clay or rock) then the loss of lateral support can lead to instability (buckling) of the pile. (Bhattacharya *et al.*, 2004). This is examined in Sec. 3.6. Finally, in Sec. 3.7, the transition from bearing failure to unstable collapse will be examined for floating pile groups, which may suffer from both effects. This will be presented in the form of charts that may be used in preliminary pile sizing to indicate the degree of excess pore pressure rise required to induce axial pile failure for various pile geometries and constructions.

It is commonly assumed that the FOS employed in selecting axial pile loads during static design will be sufficient to ensure that axial failure will not occur during liquefaction. However, due to the high degree of uncertainty involved in making this assumption, a high FOS may be used during static design. This can lead to foundations that, for most of their design life (i.e. when not subjected to earthquake shaking), are very inefficient. By studying the various possible axial pile failure mechanisms in liquefiable soil (detailed above), it will be possible to make more deterministic estimates of suitable FOS, thereby reducing the uncertainty currently inherent in the selection of this important parameter. As a result, it will be possible to design more efficient foundations in liquefaction-prone regions and allow designers to have more confidence in the suitability of their foundation solutions.

3.3 Axial Load Transfer Due to Liquefaction

3.3.1 Liquefaction-induced (co-seismic)

Axial load transfer in piled foundations during liquefaction has only recently attracted significant research interest. Rollins and Strand (2006) conducted tests on a full-scale instrumented pile. This was installed through 8m of loose saturated sand, into a bearing layer of denser sand to ~25 pile diameters (D_0). Liquefaction was induced by blasting within the loose (liquefiable) layer. The pile was observed to settle by less than 7mm ($0.02D_0$) and the shaft friction within the loose layer reduced to approximately zero as a result of liquefaction.

In the aforementioned work by Rollins and Strand (2006), no blasting was carried out in the dense bearing layer. As a result, the underlying dense sand-bearing layer in which the pile was founded did not suffer a significant increase in pore pressure and was therefore able to mobilise sufficient resistance to prevent excessive settlement of the pile. As mentioned in Sec. 3.1, although dense sands may not liquefy, significant excess pore pressures can be generated. Knappett (2006) and Knappett and Madabhushi (2008a) report dynamic centrifuge test results for 2×2 instrumented pile groups passing through loose liquefiable sand and bearing in dense sand. Liquefaction was induced by sinusoidal (harmonic) base shaking allowing for the development of excess pore pressures within the bearing layer. The resulting load transfer occurring during one of these tests is shown in Fig. 3.3.

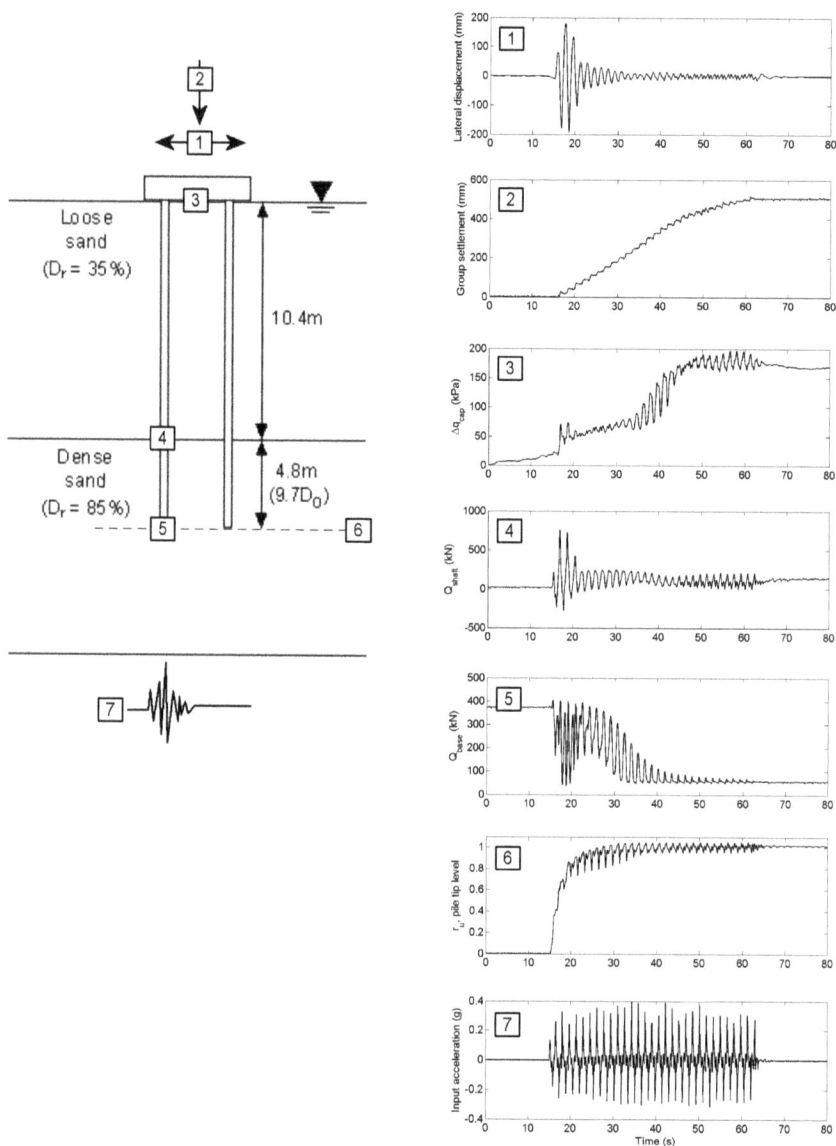

Fig. 3.3 Changes in load carried by piles and pile caps for foundations in which significant excess pore pressure is generated at pile tip level.

Figure 3.3 shows the response of the pile group (in terms of the cyclic lateral displacement and group settlement of the foundation) along with changes in the bearing pressure on the underside of the pile cap (Δq_{cap}) and changes in shaft (Q_{shaft}) and base load (Q_{base}) within one of the piles. From Fig. 3.3, the following conclusions can be drawn.

- As excess pore pressures at pile tip level increase, the load carried by the pile tip reduces. As positive settlement is occurring this may be interpreted as a drop in bearing capacity.
- If the base capacity reduces, giving rise to large co-seismic group settlement, shaft capacity can be mobilised and *increase* during the earthquake. It is currently thought that this is enabled due to shearing-induced dilation at the interface between the shaft and the sand, predominantly in the dense (bearing) layer.
- Pile group settlement increases in a stepwise manner and is correlated with the cycles of strong shaking. This suggests that the settlement is accrued from small inelastic settlements during each cycle. This would suggest that the process is governed by cyclic mobility in the dense sand rather than the loss of shaft friction due to liquefaction within the loose layer.
- The pile cap bearing capacity increases with settlement. This is due to shear-induced reduction in excess pore pressure in the soil immediately beneath the cap and plays an important role in reducing pile group settlement (see Sec. 3.4.1).

From these observations, it is clear that liquefaction-induced bearing capacity failure may occur if the excess pore pressures become sufficiently high in the bearing layer. Additionally, at lower values of excess pore pressure, liquefaction-induced settlement may still be excessive for certain types of structure (e.g. bridges) so that this may need to be accounted for in the selection of a suitable FOS for a piled foundation. Methods for including these effects in design will be discussed further in Secs. 3.4 and 3.5.

3.3.2 Downdrag (post-earthquake)

Following the end of seismic shaking, liquefied soil will reconsolidate; a process which starts at the bottom of the liquefied material and moves upwards (c.f. Sec. 3.1.1). The reconsolidation and settlement of liquefied sand will impose downwards shaft friction on the section of the pile shaft within the liquefiable layer. In addition, any nonliquefiable layers sitting on top of the liquefied sand will also be dragged down and impose additional downward loads on the piles.

In the aforementioned research of Rollins and Strand (2006) and Knappett and Madabhushi (2008a) load transfer within the instrumented piles was measured during the dissipation of excess pore pressures (reconsolidation). As the piles in each case had different outside diameters and lengths within the liquefiable layer, the measured downdrag loads have been normalised by the pile shaft surface area to obtain an average skin friction value. These are presented in Table 3.1.

Table 3.1 Downdrag loads imparted during reconsolidation of liquefied soil.

Source	Measured downdrag (kN)	Shaft length in liquefiable layer (m)	Pile diameter (m)	Average change in skin friction (kPa)
Knappett (2006)	104.5	10.4	0.496	-6.5
Rollins and Strand (2006)	80	8.0	0.324	-9.8

3.4 Pile Settlement

3.4.1 Liquefaction-induced (co-seismic)

The settlement of piles in liquefying soil has been studied by De Alba (1983), Knappett (2006) and Knappett and Madabhushi (2008a). The former of these investigations involved examining model piles within a calibration chamber. All investigations showed increased settlement with increasing r_u at the base of the pile ($r_{u,base}$). In the latter work, the use of centrifuge modelling allowed both the axial and lateral response of pile groups to be examined simultaneously.

Fig. 3.4 shows the lateral and axial response of a 2 × 2 group of long piles as shown. The two graphs in Fig. 3.4 show the response at two different values of pile axial load (represented by FOS). At low values of $r_{u,base}$, the lateral inertial and kinematic forces on the pile dominate and the peak cyclic displacements are much larger than pile settlements. As $r_{u,base}$ increases, lateral displacements of the pile reduce while settlement increases. It can also be seen from Fig. 3.4 that as the FOS reduces, significant settlement occurs at a lower value of $r_{u,base}$.

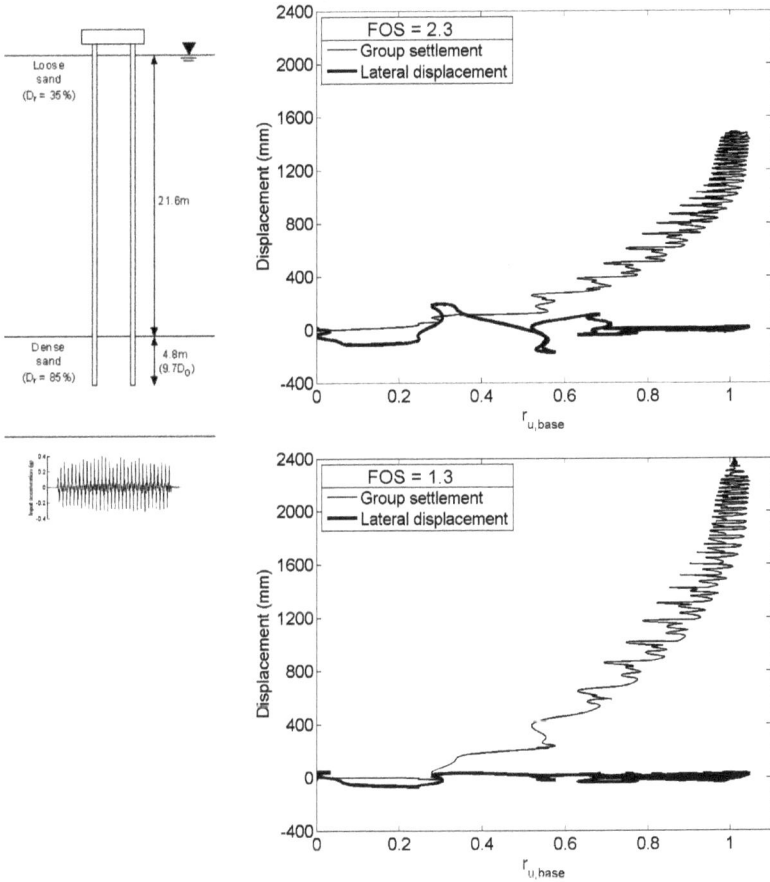

Fig. 3.4 Relative magnitudes of settlement and cyclic lateral displacement with increase in excess pore pressure ratio at pile tip level.

It is clear, therefore, that even a partial increase in pore pressure in the bearing layer can lead to damagingly large settlements – it is not an essential condition for $r_{u,base} = 1$ to be reached. Despite measurements of pile settlement having been made during laboratory experiments, the complex interactions between the pile cap, shaft and base make difficult the deterministic prediction of pile settlement due to seismic liquefaction. Empirical methods, however, seem to suggest that for a given limiting settlement, the minimum FOS required can be directly related to $r_{u,base}$. Fig. 3.5 shows collated data from the work of both De Alba (1983) and Knappett (2006). A line can be empirically fitted through these data of the form

$$FOS = 1 + A(r_{u,base})^B. \qquad (3.2)$$

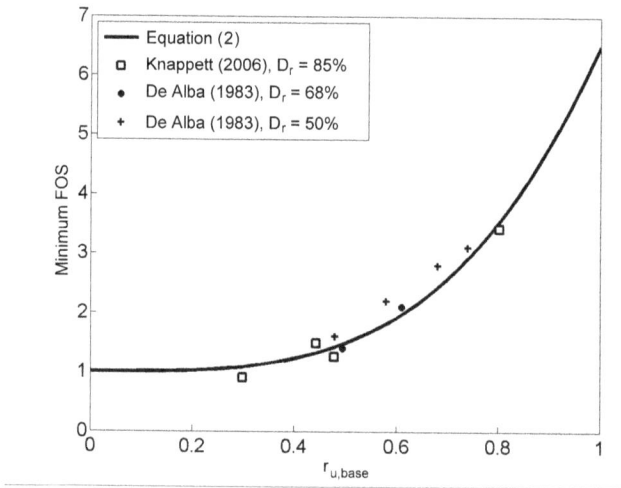

Fig. 3.5 Minimum FOS for a given degree of liquefaction at pile tip level for end bearing piles, showing insensitivity to soil density.

For a limiting pile settlement of $0.1D_0$, as shown in Fig. 3.5, A = 5.5 and B = 3.5. Figure 3.5 shows that the relationship between FOS and $r_{u,base}$ is relatively insensitive to the density of the bearing layer. De Alba (1983) further showed that this relationship is also insensitive to confining stress (or, alternatively, pile tip depth). This is to be expected as it is implicitly incorporated into the FOS. In the absence of a more deterministic solution, Fig. 3.5 or Eq. 3.2 may be used to select a suitable axial load (or FOS) if serviceability criteria are to be met in seismic pile design (e.g. for an L1-class earthquake).

As previously mentioned in Sec. 3.3.1, the pile cap also has an important role to play in reducing the settlement of pile groups. Figure 3.6 shows a comparison between two 2 × 2 pile groups of 15.2m length passing through a loose layer of saturated sand 10.8m thick and bearing in dense sand. Both pile groups carry the same total vertical load, but vary in terms of the contact between the underside of the pile cap and the upper surface of the loose sand. In the figure, the 'seismically-induced axial load' is the additional cyclic component that is imparted at the pile head due to foundation rocking during the earthquake (i.e. the monotonic component due to the dead weight of the supported structure has been filtered out).

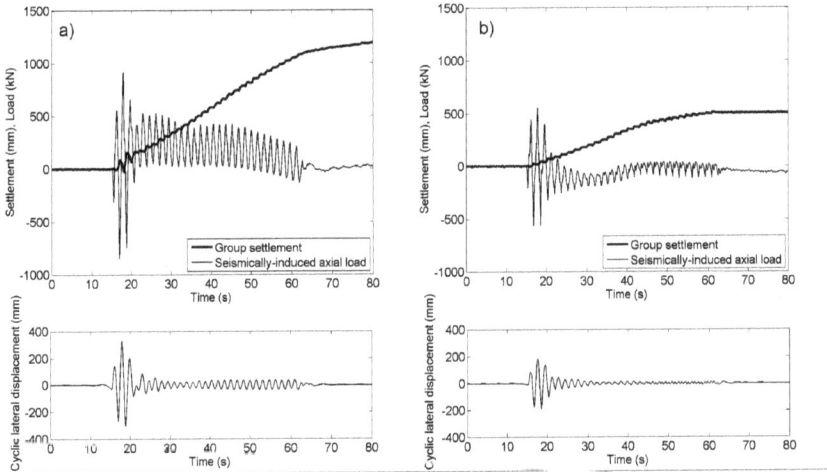

Fig. 3.6 Group settlement, seismically-induced axial pile load and lateral cyclic response of 2 × 2 pile groups of identical configuration (a) without the pile cap contributing to group behaviour and (b) including the effect of the pile cap resting on the surface of loose liquefied sand.

Two important conclusions can be drawn from Fig. 3.6.

- The cap is able to contribute bearing resistance to the pile group, which leads to lower group settlement. This is because the cap is able to 'take up' some of the load shed from the pile tips as the bearing capacity of the bearing layer reduces.
- The cap provides additional rotational stiffness to the pile group and thereby reduces both the additional axial load that is induced in the measured pile during shaking and also the lateral displacements (and hence bending moments and shear forces) induced in the piles.

The first of these points is particularly important as it implies that if the piles are designed neglecting the effects of the pile cap (see Sec. 3.5), then the cap will provide an intrinsic additional FOS against failure.

3.4.2 *Downdrag (post-earthquake)*

Settlements occurring due to load transfer during reconsolidation of liquefied sand (c.f. Sec. 3.3.2) may be estimated following the method presented by Boulanger *et al.* (2003). This method is essentially a modification of the neutral plane solution originally developed by Fellenius (1972). To compute the settlement at a time t_1 after shaking has stopped and the soil is fully liquefied (time t_0) requires the isochrone of excess pore pressure at that time to be estimated. Once this has been done, analysis follows a three-stage process for each time increment $(t_0 \rightarrow t_1)$:

- Compute soil settlement profile due to reconsolidation increment.
- Determine neutral plane location.
- Pile settlement increment is value of soil settlement increment at neutral plane depth (from definition of the neutral plane).

This process is shown schematically in Fig. 3.7.

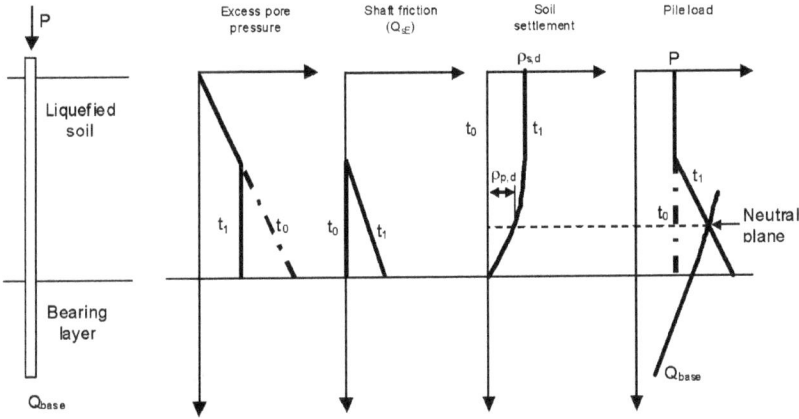

Fig. 3.7 Computation of downdrag-induced settlements using neutral plane solution of Boulanger et al. (2003).

In step one, soil settlements are found by integrating the vertical strains (ε_v) in the liquefiable layer, which are in turn found by integrating the product of the vertical effective stress change and soil compressibility of the reconsolidating sand, as shown in Eq. 3.3

$$\varepsilon_v = \int_{t_0}^{t} \sigma'_v m_v dt . \quad (3.3)$$

where

$$m_v = m_{v0} \cdot \frac{exp(\xi)}{1 + \xi + 0.5\xi^2} \quad (3.4)$$

$$\xi = 5(1.5 - D_r)r_u^b \quad (3.5)$$

$$b = 3(4^{-D_r}) \quad (3.6)$$

(Seed *et al.*, 1975). In Eqs. 3.5 to 3.6, D_r is the relative density of the liquefiable soil and m_{v0} is the soil compressibility in the absence of any excess pore pressures (i.e. $r_u = 0$). The soil around the pile is assumed to reconsolidate so that shaft friction within soil for which $r_u < 1$ is assumed to be given by

$$Q_{sE} = \sigma'_{v0}(1 - r_u)K_0 \tan\delta \quad (3.7)$$

where δ is the interface friction angle and K_0 the coefficient of lateral earth pressure at rest.

The neutral plane is the depth above which relative soil-pile settlements are positive and below which, soil-pile settlements are negative. The depth of the neutral plane (step two) is calculated by summing the loads on the pile from top to bottom, starting at the pile head load P and assuming that values of Q_{SE} from Eq. 3.7 act *downwards* on the pile, then summing the loads from bottom to top, assuming that Q_{SE} within the reconsolidating sand acts *upwards* on the pile. The point at which these two lines intersect gives the location of the neutral plane. At this depth, soil and pile settlements are equal, so that the pile settlement at a given time may be found by interpolation from the soil settlements computed in step one.

The example shown in Fig. 3.7 considers a two-layer soil profile of liquefied soil overlying a more competent bearing layer. The presence of nonliquefiable crustal layers on top of the liquefiable material may also be accounted for by the above method. This is presented in further detail by Boulanger *et al.* (2003). However, it is important to know that this case can be more critical as the crustal layers will suffer the same (maximum) settlement of the top of the reconsolidating soil beneath, while applying large frictional forces to the pile. It should also be noted that if the bearing layer is cohesionless and saturated, excess pore pressures may be generated, lowering the loads carried at the base of the pile. This should be taken into account during step two (see Eq. 3.8 in the following section).

3.5 Guidelines for Designing Against Bearing Failure

It was shown in Sec. 3.3.1 that the development of excess pore pressures within dense bearing layers leads to a reduction in pile tip bearing capacity. This is likely to be significant for L2-class earthquakes involving stronger seismic shaking. It would therefore be beneficial to be able to predict quantitatively how bearing capacity will reduce as the excess pore pressure ratio in the soil changes. Ideally, this would provide a relationship between limiting minimum FOS and $r_{u,base}$ with bearing failure as the limiting condition, similar to Eq. 3.2 for a settlement-based limiting condition.

By considering a modified spherical cavity expansion solution for pile base capacity originally proposed by Vesic (1972), and considering that the effect of the excess pore pressure rise is to reduce the effective stress around the pile tip, it may be shown (Knappett and Madabhushi, 2008b) that the pile base capacity during earthquake shaking in saturated soil ($Q_{base,E}$) is related to the initial static ultimate base load ($Q_{base,S}$) by

$$\frac{Q_{base,E}}{Q_{base,S}} = \left(1 - r_u\right)^{\frac{3-\sin\phi}{3(1+\sin\phi)}} . \tag{3.8}$$

where ϕ is the angle of friction of the soil beneath the pile tip. Figure 3.8 shows the displacement response of the pile group discussed in Fig. 3.6, along with base load measured at the tip of one of the piles and the base capacity, computed using Eq. 3.8. This shows a correlation between the onset of large settlements (the change in the gradient in the top region of the figure) and the base capacity reducing below the value of the applied pile tip load. This suggests that bearing capacity failure may be avoided by ensuring that, for a given value of $r_{u,base}$, the initial FOS for a given pile is selected so that the initial base load carried by the pile (Q_{b0}) is less than the minimum base capacity given by Eq. 3.8, i.e.

$$Q_{b0} \leq Q_{base,E} . \tag{3.9}$$

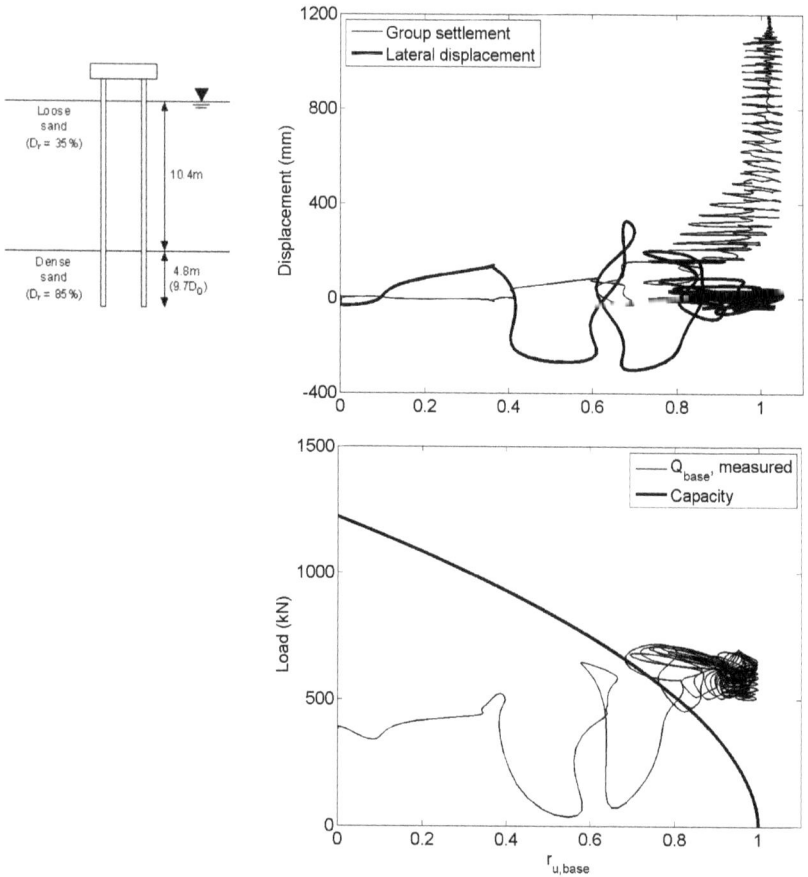

Fig. 3.8 Onset of large pile group settlements as pile base load exceeds capacity as given by Eq. 3.2.

By combining Eqs. 3.8 and 3.9 and considering the distribution of load (and capacity) between the base and shaft of piles in sandy soil, this leads to

$$FOS \geq \frac{1}{\alpha_{ult}\left(1-r_u\right)^{\frac{3-\sin\phi}{3(1+\sin\phi)}} - \alpha_{ult} + 1} \tag{3.10}$$

where α_{ult} is the ratio of the static base capacity to the total static pile capacity, both of which are calculated when undertaking conventional static pile design. Eq. 3.10 has been plotted in Fig. 3.9 for common values of ϕ and α_{ult}, and shows the minimum value of the FOS below which liquefaction-induced bearing capacity failure would be expected. The limiting settlement values given by Eq. 3.2 (Sec. 3.4.1) are also shown for comparison.

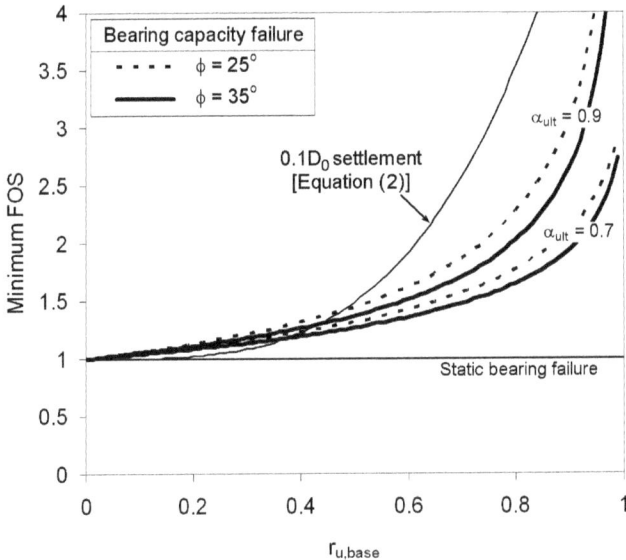

Fig. 3.9 Design chart to select a suitable FOS during pile design to avoid liquefaction-induced bearing capacity failure (Eq. 3.4).

3.6 Instability of Single Piles and Pile Groups

3.6.1 Rock-socketed piles

As detailed in Sec. 3.2, the loss of lateral soil support to a pile due to liquefaction can cause a dramatic drop in the lateral stiffness of a piled foundation. Liquefaction-induced instability has recently been investigated by Bhattacharya (2003) and Bhattacharya *et al.* (2004) who considered single piles and Knappett (2006) investigating 2 × 2 pile groups. In both series of tests the piles were modelled as rock-socketed to avoid the possibility of bearing failure so that instability could be studied in isolation.

Considering the simplest model of liquefied soil as a material having zero strength and stiffness, a pile passing through a fully liquefied stratum of soil is expected to behave similarly to an unsupported column. Under these conditions, the axial critical load at which buckling occurs is given by Euler's equation

$$P_{cr} = \frac{\pi^2 EI}{\left(\beta L_p\right)^2} \qquad (3.11)$$

where β is a factor accounting for the fixity at both ends of the pile. Values of β for common fixity conditions are shown in Fig. 3.10. Essentially, βL_p is the equivalent length of a pinned-pinned strut which has the same critical load as a strut with the given fixity conditions.

Using Fig. 3.10 in combination with the analogy of piles surrounded by liquefiable soil as unsupported columns, for single cantilever piles (fully fixed at the base by the rock socket), $\beta = 2$, while for piles which are part of a group (i.e. fixed at the top by the pile cap, but free to sway) $\beta = 1$. βL_p is also termed the effective length of the pile, and can be normalised by the radius of gyration of the pile section to give the slenderness ratio, $\beta L_p/r_g$. At low values of slenderness ratio, critical loads are much higher than plastic squashing loads so that the latter effect controls failure; such piles are considered 'stocky'. Piles that fail by buckling rather than squashing are 'slender' and have higher values of slenderness ratio. Figure 3.11 shows a summary of the results of dynamic centrifuge tests on both single piles and piles in groups carrying axial loads in fully-liquefied soil, which clearly shows that such piles will buckle if P_{cr} is exceeded. A poor performance relates to the pile(s) suffering sudden catastrophic failure.

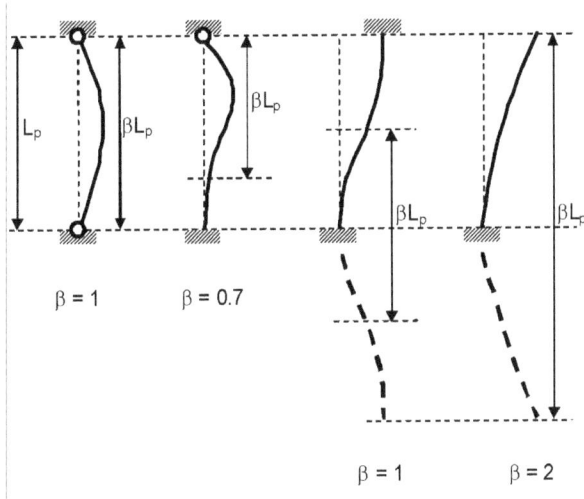

Fig. 3.10 Fixity factors (β) for some common fixity conditions.

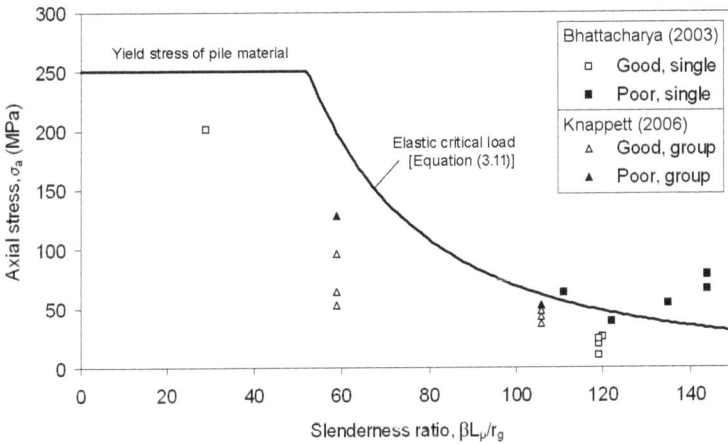

Fig. 3.11 Summary of dynamic centrifuge test results on rock-socketed piles in fully liquefied soil.

It is also apparent from Fig. 3.11, however, that piles can become unstable at loads below critical. The reason for this is that piles in level ground will often have small imperfections from vertical due to lateral displacement at the pile head, which may occur cyclically due to earthquake shaking. The presence of axial load will cause amplification of these lateral displacements. If the amplified displacement creates sufficient curvature in the pile to cause yield, the pile will collapse at an axial load lower than critical. Continuing the analogy from earlier in this section, the effect of imperfections on axial failure loads can be expressed in terms of the Perry–Robertson Equation:

$$\left(\sigma_a\right)^2 - \left[\sigma_y + \left(1 + \frac{\Delta_0 D_0}{2r_g^2}\right)\frac{P_{cr}}{A_p}\right]\sigma_a + \sigma_y \frac{P_{cr}}{A_p} = 0 .\tag{3.12}$$

where σ_y is the column yield stress, Δ_0 is the size of the imperfection, D_0 is the pile diameter, A_p the cross-sectional area of the column and r_g and P_{cr} are as defined previously. Eq. 3.12 plots as a continuous curve on axes of σ_a versus slenderness ratio (i.e. as in Fig. 3.12).

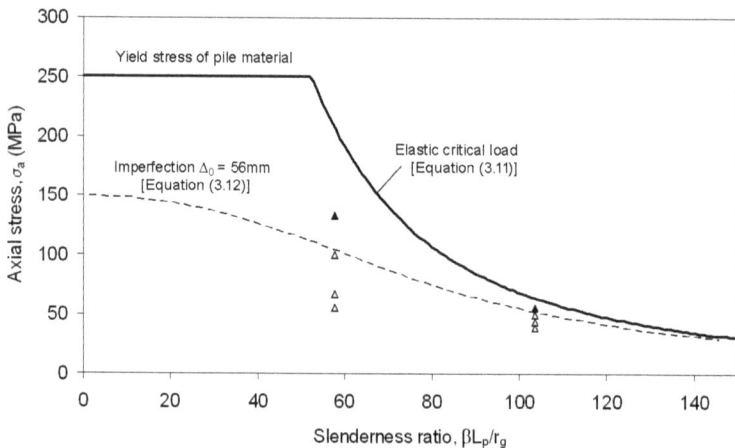

Fig. 3.12 Comparisons of computed behaviour for perfect (Eq. 3.11) and imperfect (Eq. 3.12) piles with experimental data.

For increasing imperfection size, the maximum stress that can be sustained before collapse occurs reduces for a given slenderness ratio. In the centrifuge tests of Knappett (2006), the size of imperfections in the pile groups before liquefaction occurred was measured and the data collected has been compared to Eq. 3.12 for a typical value of Δ_0 measured during the tests. This is shown in Figure 3.12, which shows close agreement between the experimental data and Eq. 3.12.

As well as reducing the axial collapse load, initial imperfections may be amplified (increased) due to the axial loads present in the piles. Amplification of lateral displacements has additionally been studied for unsupported columns by Timoshenko and Gere (1961) in which the amplification factor is related to the fraction of critical load carried according to

$$Amplification = \frac{1}{1-\frac{P}{P_{cr}}} \,. \tag{3.13}$$

In the dynamic centrifuge tests reported previously by Knappett (2006), lateral deflections were measured at the pile heads before, during and after liquefaction. Comparison between these results and Eq. 3.13 are shown in Fig. 3.13. It can be seen that the match between observed and predicted behaviour becomes worse as the piles become more slender. For stockier piles, the pile flexibility will tend to dominate compared to that of the soil, while for more slender piles, the soil will have a much greater influence on the overall response. This would suggest that the differences between the two sets of results is due to the stiffness of the liquefied soil being underestimated by considering it to be zero. Improved estimates of soil stiffness would therefore be expected to improve the match at high slenderness ratio while having a smaller effect on the better match at lower slenderness.

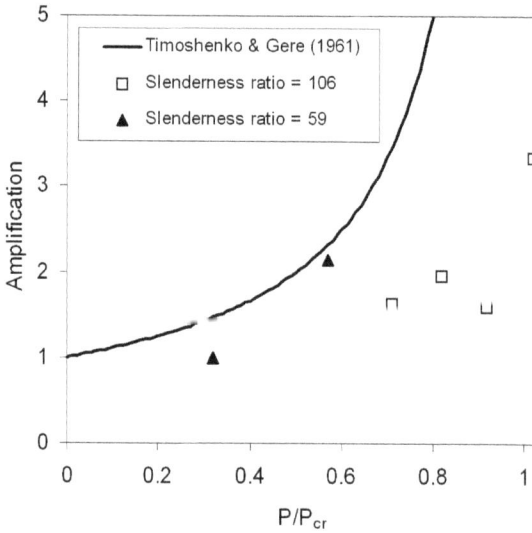

Fig. 3.13 Comparison of measured amplifications from dynamic centrifuge tests against predictions using a zero strength and stiffness soil model, Eq. 3.13.

In the case of both single piles and pile groups, the onset of instability leads to collapse due to the formation of plastic hinges in the piles. For soil liquefied over its full depth, the locations of maximum bending moment (and therefore also plastic hinges) would be expected to occur at the bottom of the piles socketed into underlying firm ground (e.g. bedrock), and immediately beneath the cap (in the case of a pile group). However, observations of collapsed piles and pile groups from centrifuge model tests revealed that while hinges did form beneath the pile cap, the lower set of hinges occurred at an intermediate depth in the piles. Bhattacharya (2003) proposed that this was due to resistance developed within the liquefied soil as the pile shears the soil. Simplifying the soil reaction as a constant value with depth which is independent of relative soil-pile displacement gives the following equation for the depth below pile head level at which the pile hinge would form in a cantilever (free-head) pile:

$$z_h = 1.165\sqrt{\frac{EI}{P}}$$ (3.14)

where P = axial pile load and EI = pile bending stiffness. Eq. 3.14 was found to predict hinges at an intermediate depth within a single free-head pile and gave a reasonable match to observed hinge locations.

Detailed measurements of the response of liquefied soil to large monotonic relative soil-pile displacement by Takahashi *et al.* (2002) and Towhata *et al.* (1999) have revealed that the lateral soil-pile reaction is highly nonlinear with lateral displacement. At low lateral soil-pile displacement, the resistance of the soil is at a low residual value. After a certain displacement however, the local excess pore pressures around the pile begin to drop sharply, leading to a consequent increase in soil-pile resistance. Takahashi *et al.* (2002) termed this displacement (when normalised by pile diameter) the reference strain of transformation (γ_L) as it qualitatively indicates a phase transformation point at which the soil switches from contractile to dilative behaviour. They further found that its value is heavily dependent on pile shearing velocity, with γ_L reducing as shearing velocity increases. γ_L also increases with depth corresponding to suppression of dilative behaviour under increasing total stress. The lateral soil-pile behaviour during large displacement events (e.g. the post-buckling behaviour of a pile) is shown schematically in Fig. 3.14.

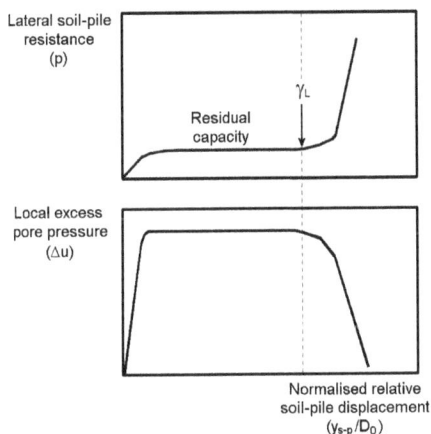

Fig. 3.14 Schematic representation of liquefied soil response to soil-pile relative displacement.

In modelling the response of pile groups, Knappett (2006) undertook numerical modelling using a beam on nonlinear Winkler foundation (BNWF) model so that nonlinear lateral soil-pile behaviour of the form shown in Fig. 3.14 could be modelled. These models were analysed using a Riks Algorithm, which can model the full post-buckling response (plastic collapse) of pile groups as the axial critical load is approached. (Riks, 1972; 1979). Prior to collapse, the amplification of lateral deflections can also be examined within the same model. Figure 3.15 shows a comparison between the collapsed mode shape and pile stresses from the numerical model with the corresponding mode shape and observed locations of plasticity from the corresponding centrifuge model. The numerical model is also able to give improved predictions of amplification, as shown in Fig. 3.16. This confirms that it is the liquefied soil response which is responsible for the mismatch between predicted and measured amplification shown in Fig. 3.13.

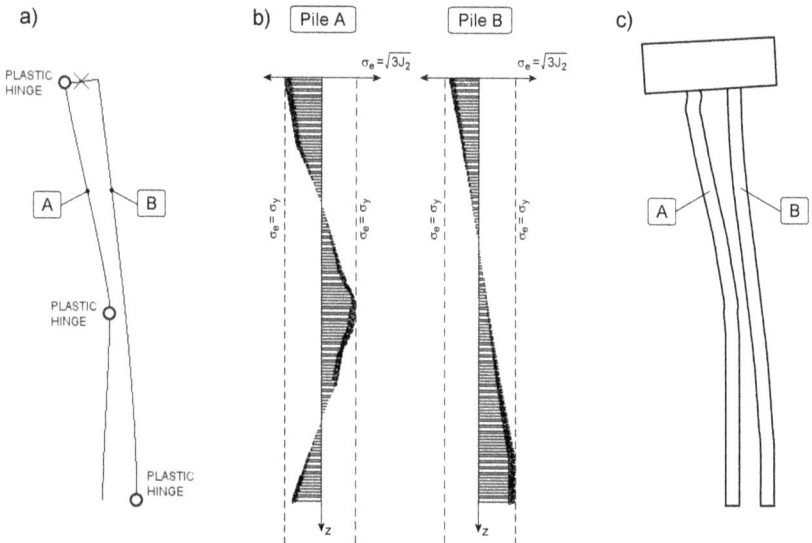

Fig. 3.15 Comparisons of numerical post-buckling simulations with centrifuge test results: (a) collapse mechanism from simulation; (b) Von-Mises stresses within piles showing yield locations; (c) permanent (plastic) deformations of centrifuge model measured post-test (displacements at exaggerated horizontal scale).

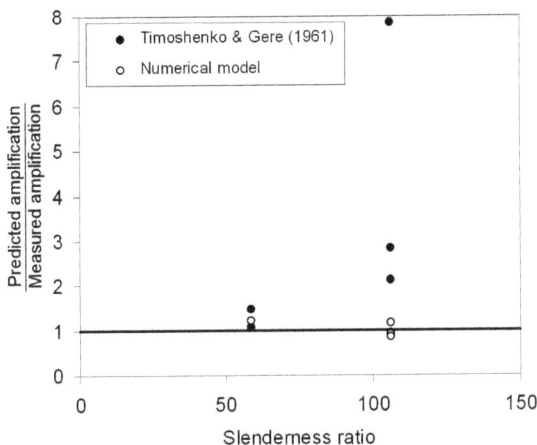

Fig. 3.16 Amplification of lateral displacements due to axial load acting on 2 × 2 pile groups in fully liquefied soil, showing the improvement in prediction using the soil-pile interaction behaviour shown in Fig. 3.14 compared to simple predictions using Eq. 3.13.

3.6.2 Floating piles

The previous section (3.6.1) demonstrated how rock-socketed piles, i.e. piles with practically infinite bearing resistance, can buckle during liquefaction. For pile groups consisting of floating piles, there is additionally the possibility of failure in bearing as outlined in Sec. 3.5. As the bearing resistance varies with the depth of the liquefaction front (expressed in terms of $r_{u,base}$), it is useful to be able to determine the critical buckling loads also in terms of a variable depth of liquefaction front. This may be accomplished analytically by considering the lateral resistance that the soil is able to provide as liquefaction progresses.

Figure 3.17 details schematically the vertical effective stress conditions within the soil around a floating pile group as the liquefaction front progresses to a depth (z_L). A simple bilinear approximation to the excess pore pressure distribution is made here in which soil shallower than z_L is at full liquefaction ($r_u = 1$ or $\sigma'_v = 0$), and the excess pore pressure is constant with depth below z_L. The resulting effective stress distribution is identical to that for partially embedded piles with a free-standing length of z_L.

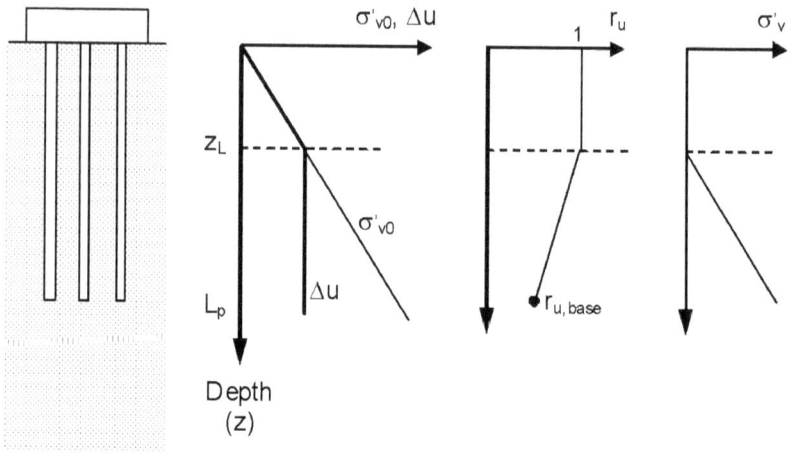

Fig. 3.17 Effective stress conditions around piles for a given depth of full liquefaction (z_L).

Gabr *et al.* (1997) present an analytical solution for determining the elastic critical loads for partially embedded piles; a brief summary of the procedure is given here. This method is able to account for different distributions of horizontal subgrade reaction, k (i.e. k = constant or k = $\eta_h z$, where η_h is the modulus of subgrade reaction), axial pile load distribution (split between base and shaft resistance) and boundary conditions at the top and tip of the piles. Due to the continuous nature of the piles, there are an infinite number of degrees of freedom. Critical loads are therefore found using the Rayleigh–Ritz method, which considers the stored energy due to combined flexure of the pile and elastic deformation of the soil (U) and the potential energy of the external axial load (V) for a given lateral pile displacement, $y_p(x)$. The critical load and corresponding mode shape are then found, which minimise the sum of these two energy terms. The terms U and V are found from Eqs. 3.15 and 3.16.

$$U = \frac{EI}{2} \int_0^{L_p} \left(\frac{d^2 y_p}{dx^2} \right)^2 dx + \frac{D_0}{2} \int_0^{L_p - z_L} p(x) y_p dx \qquad (3.15)$$

$$V = -\frac{1}{2}\int_0^{L_p} P(x)\left(\frac{dy_p}{dx}\right)^2 dx \qquad (3.16)$$

In Eqs. 3.15 and 3.16, x = L_p − z (i.e. a coordinate measured upwards from the tip of the pile) and p(x) is the lateral soil-pile reaction at a given height above the pile tip. For piles in cohesionless liquefiable sands in which k = $\eta_h z$

$$p(x) = \eta_h\left(L_p - z_L - x\right)y_p(x). \qquad (3.17)$$

It should be noted in Eq. 3.17 that the displacement of the surrounding soil is assumed to be negligible so that the relative pile-soil displacement is given by $y_p(x)$ alone. In order to solve Eqs 3.15–3.17, the function used to represent the lateral displacement of the pile $y_p(x)$ must satisfy the boundary conditions at the top and tip of the pile, even though the exact shape between these extremes is not yet known. This is achieved by using a 'shape function', which is split into a number of different Fourier components, which are superimposed by varying amounts to give $y_p(x)$. In the case of floating piles, within a pile group, that are fixed with sway at the top and free at the tip, the shape function is given by

$$y(x) = c_0 + \sum_{n=1}^{N} c_n \sin\frac{2n-1}{2L_p}\pi x. \qquad (3.18)$$

The constants c_n in Eq. 3.18 are found during the analysis to determine the exact mode shape. As the number of Fourier terms (N) is large, $y_p(x)$ becomes an increasingly better representation of the actual mode shape. The form of Eq. 3.18 ensures that the boundary conditions are met irrespective of the values of the constants c_n and N. Determination of the critical loads is accomplished by substituting the shape function into Eqs. 3.15 and 3.16 and applying the principle of minimum potential energy, i.e.

$$\frac{\partial(U+V)}{\partial c_n} = 0. \qquad (3.19)$$

This leads to a set of N+1 homogeneous linear equations in terms of c_n, which may be written in matrix form as

$$[A]\underline{c}_n = 0.$$

(3.20)

For the pile fixity conditions considered herein (for a pile group), the direct formulation of matrix [A] may be found in Gabr *et al.* (1997) without the need to undertake the minimisation procedure represented by Eq. 3.19. The system of Eqs. 3.20 has nontrivial solutions for $\underline{c}_n$ only if the determinant of the matrix [A] is zero This is a statement of the generalised eigenvalue problem which can be solved by standard methods. The lowest eigenvalue (λ_1) is related to the effective length factor introduced in Sec. 3.6.1 by

$$\lambda_1 = \frac{1}{\beta^2}$$

(3.21)

and can be used to find the critical load by substitution of the resulting value of β in Euler's Eq. 3.11. The method described in this section for floating pile groups will be used to find elastic critical loads for various pile configurations in the following section.

3.7 Bearing vs. Buckling Failure

3.7.1 Methodology

For floating pile groups, the failure mechanisms of bearing (settlement) failure and buckling will both be possible. In this section, the limiting axial loads for each mechanism will be compared for various pile geometries and material properties to determine their vulnerability to failure in the two modes. The properties that will be considered here include:

- pile length (L_p);
- pile diameter (D_0);
- pile material/construction (EI);
- soil density (η_h, $Q_{base,S}$).

The ultimate bearing capacity P_{bc} of a pile as a function of liquefaction propagation was determined assuming that the initial static base capacity reduces according to Eq. 3.8, while the shaft capacity is dependent on the depth of the pile below the liquefaction front.

$$P_{bc} = Q_{base,S} \left(1 - r_{u,base}\right)^{\frac{3-\sin\phi}{3(1+\sin\phi)}} + Q_{shaft,S} \frac{L_p - z_L}{L_p}. \qquad (3.22)$$

Referring to Fig. 3.17, it can be seen that

$$r_{u,base} = \frac{z_L}{L_p} \qquad (3.23)$$

so that the change in bearing capacity due to liquefaction propagation (Eq. 3.22) can be expressed solely in terms of $r_{u,base}$. $Q_{base,S}$ is found herein according to the method proposed originally by Berezantzev (1961), as modified by Cheng (2004). $Q_{shaft,S}$ was calculated assuming an interface friction angle $\delta = \phi/2$ according to

$$Q_{shaft,S} = \pi D_0 \int_0^{L_p} \sigma'_{v0} K \tan\left(\delta\right) dz \qquad (3.24)$$

where $K = K_0 = 1 - \sin\phi$. For use in design, $Q_{base,S}$ and $Q_{shaft,S}$ may alternatively be determined from CPT data as detailed in Sec.1.1.2 in Chapter 1.

Elastic critical loads (P_{cr}) were found using the method outlined in Sec. 3.6.2. Values of β were found for a given depth of full liquefaction (or, alternatively, $r_{u,base}$), so that critical loads could be found from

$$P_{cr} = \frac{\pi^2 EI}{\left[\beta\left(r_{u,base}\right)L_p\right]^2}. \qquad (3.25)$$

For a given pile geometry, initial static axial pile loads were compared to the capacities P_{bc} and P_{cr}, allowing the value of $r_{u,base}$ at which either of the capacities was exceeded to be determined. The results are presented here in the form of charts that plot $r_{u,base}$ against pile length (L_p) for a given pile section (D_0 and EI) and FOS (axial load).

3.7.2 Sample analysis

For a given pile material and construction, the cross-sectional properties (D_0 and EI) will be related. For steel tubular piles, this relationship is given by

$$EI = E_s \frac{\pi}{4}\left[\left(\frac{D_0}{2}\right)^4 - \left(\frac{D_0}{2} - t\right)^4\right]$$ (3.26)

where E_s is the Young's modulus of the steel and t is the wall thickness. For solid, reinforced concrete (RC) piles, the expression is slightly more complicated as:

- the pile is made of two dissimilar materials (concrete and steel);
- the designer is free to choose the percentage of steel that is used in the pile;
- concrete can only effectively carry load in compression.

Nevertheless, EI for a solid circular RC pile may be found using transformed area theory. (Kong and Evans, 1987). Considering the neutral axis to be at the centre of the beam and assuming that the reinforcing steel is uniformly distributed as an annulus of material with a given cover depth (c) within the concrete gives

$$EI = E_c\left\{0.0245D_0^4 + \frac{\alpha A_s}{4}\left[2\left(\frac{D_0}{2} - c\right)^2 - \frac{\alpha A_s}{\pi}\right]\right\}$$ (3.27)

where E_c is the Young's modulus of the concrete and α is the modular ratio (= E_s/E_c). It is common in pile analysis to model concrete piles by assuming elastic behaviour of a solid section of concrete, instead of using Eq. 3.27. While this requires fewer parameters, it has recently been shown by Knappett (2008) that such a simplification can grossly overestimate EI, particularly for piles with lower proportions of reinforcing steel.

For RC piles with A_s/A_c = 3%, representative of an RC pile designed to resist bending, and steel tubular piles with t = 16mm, Figure 3.18 shows the variation of EI with D_0.

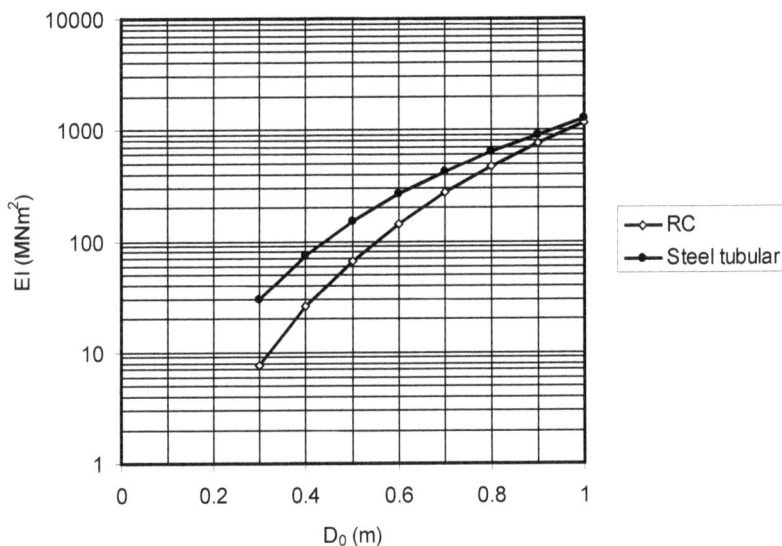

Fig. 3.18 Section properties of steel tubular piles (t = 16mm) and solid circular RC piles (A_s/A_c = 3%).

Figure 3.19 shows the results of the analyses for piles designed to a static FOS = 2 resting in a single layer of loose saturated sand with the following properties: $D_r \approx 35\%$, $\phi = 32°$, $\eta_h = 6.6\text{MN/m}^3$.

3.7.3 Ultimate axial limiting states for piled foundations

From Fig. 3.19 it is possible to draw the following general conclusions relating to the axial design of piles in level liquefiable soils.

- For short piles with low bearing capacity and low effective lengths, the dominant mode of failure is bearing. As the pile gets longer, so the critical value of $r_{u,base}$ required for failure reduces.
- At longer pile lengths, instability becomes the dominant mode of failure. The onset of buckling can happen at very low values of $r_{u,base}$; however, r_u is likely to be lower at these depths for a given earthquake due to the higher total stress.

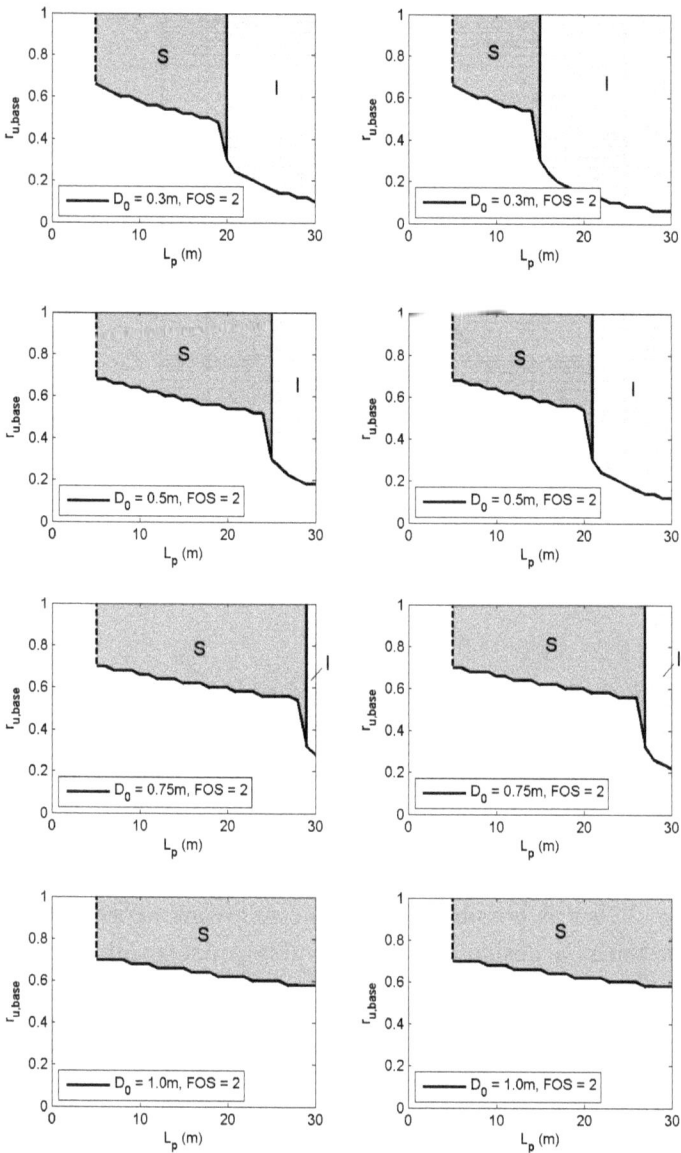

Fig. 3.19 Design charts for steel tubular piles (left column) and solid circular RC piles with $A_s/A_c = 3\%$ (right column) for piles in loose sand ($D_r = 35\%$) and static FOS = 2. 'S' denotes regions in which bearing failure will occur, while 'I' denotes instability (buckling).

- As piles reduce in size (and therefore stiffness) the transition from bearing to buckling failure occurs at lower pile lengths. RC micropiles are therefore likely to be particularly susceptible to instability failure.
- The use of fewer large diameter piles can suppress the possibility of instability failure at pile lengths of up to 30m. The axial loads on such piles can then be selected to avoid bearing capacity failure. It also appears that RC piles are just as suitable as steel tubular piles in this regard. From Fig. 3.18 it can be seen that RC piles are able to offer a similar EI to steel tubular piles when diameter becomes large and are therefore also suitable for resisting seismic lateral (bending) loads.
- At values of $r_{u,base}$ lower than the critical values shown in Fig. 3.19 the piles must still be checked against bending failure under the earthquake-induced lateral loads (c.f. Chapter 2).

The example shown in Fig. 3.19 is for a floating pile group in homogeneous loose soil. For piles that pass through liquefiable layers and bear in more competent bearing layers, the region of settlement failure at lower pile lengths will reduce in size, i.e. the values of $r_{u,base}$ at which failure occurs are likely to increase. The size of the instability region will remain unaltered, however, as the instability of the pile is dominated by the loss of soil support in the upper regions of the pile where pile displacements are largest.

In using these recommendations it must be remembered that only the ultimate limiting state is considered here. At values of $r_{u,base}$ lower than critical there may still be significant settlement and amplification of lateral pile deflections and bending moments. For the latter effect, accurate prediction relies on simulations which can account for the nonlinear effects of axial load on lateral response. However, Eq. 3.13 may be used with estimates of P_{cr} using the method outlined in Sec. 3.6.2 to provide a conservative estimate of amplification effects. Such a prediction will be more accurate for stockier piles.

3.7.4 Use of limiting states in pile sizing

The previous section has presented a method for determining the limiting values of $r_{u,base}$ at which axial pile failure will occur in liquefiable soil, and how these vary with pile length. These limiting values define the axial *capacity* of a pile in liquefiable soil. For a given depth of liquefaction (z_L), the value of $r_{u,base}$ for any pile, given the simple bi-linear approximation used herein, is also related to pile length by Eq. 3.23. This equation plots as an hyperbola on axes of $r_{u,base}$ versus L_p and represents the *demand* on pile length for a given depth of liquefaction. By overlaying such a curve on a capacity chart (e.g. Fig. 3.19) it is possible to determine suitable pile lengths to avoid axial failure modes. This will be illustrated as an example.

Consider a 0.3m diameter steel tubular pile with the properties described in Sec. 3.7.2. The capacity chart has already been determined for this pile as shown in Fig. 3.19. Two different scenarios are considered here. In the first, $z_L = 5$m; and in the second, $z_L = 10$m (these might represent behaviour at L1 and L2 earthquakes respectively). The capacity curves using Eq. 3.23 with the given values of z_L are plotted with the capacity chart for the pile in Fig. 3.20a. It can be seen that for $z_L = 5$m, a suitable pile length would be 8–21m. For $z_L = 10$m, however, the 0.3m diameter pile is not suitable as it will fail in one or other of the axial failure modes, irrespective of the length chosen.

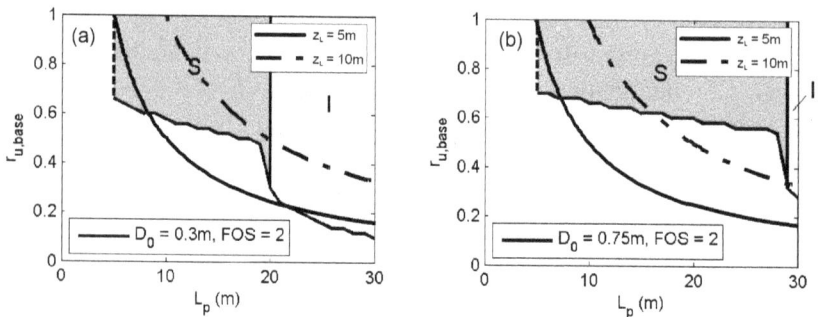

Fig. 3.20 Example of use of methods presented in Sec. 3.7 for sizing of piled foundations.

In such a case, the solution is to use a larger diameter pile. Figure 3.20b shows the same two demand curves plotted with the capacity curve for a 0.75m diameter steel pile. Increasing the diameter has made it possible to find a design pile length for z_L = 10m which will not fail (16m <L_p <28.5m).

3.8 Summary

This Chapter has introduced some of the effects that axial pile load can have on the response of piled foundations to soil liquefaction in level ground. It has been shown that bearing layers which are classed as 'nonliquefiable' by traditional methods can still generate significant excess pore pressures, which lead to pile settlement and ultimately, bearing capacity failure. Additionally, if the pile is carrying significant axial load, the loss of lateral soil support in surrounding liquefiable layers can lead to instability (buckling) of the pile. Even if collapse in this mode is avoided, the presence of axial loads will amplify lateral displacements and pile bending moments generated due to the horizontal earthquake forces (c.f. Chapter 2). The foundation and liquefaction conditions which lead to ultimate failure in both axial failure modes (bearing and buckling) have been determined and compared and the implications of this analysis for the design of piled foundations have been drawn.

Chapter 4

Lateral Spreading of Sloping Ground

4.1 Liquefaction-induced Lateral Spreading

4.1.1 Introduction

As discussed in Chapters 1 to 3, seismic shaking of loose soil deposits can lead to a build-up of excess pore pressure and hence to liquefaction of the deposit. In level ground this may cause settlement and/or rotation of structures founded upon these deposits, but in sloping ground the loss of strength involved in liquefaction may cause the slope to become unstable and to flow downhill.

This behaviour has been observed in many major earthquakes including those in Alaska in 1964, Niigata in 1964, Kobe in 1995 and Taiwan in 2001. Two examples of the damage observed in the Alaska earthquake are shown in Figs. 4.1 and 4.2. As the banks of the river move towards each other due to lateral spreading, the railway tracks passing over rivers are buckled.

Damage tends to be exerted on a variety of geotechnical structures including: wharves, bridges, shallow foundations and pile foundations. Lateral spreading is not only a problem in steeply sloping ground. In centrifuge experiments lateral spread displacements of 1m have been observed, even with a 3° ground surface slope. (Haigh et al., 2000). If the residual strength of the liquefied soil is low enough that static stability of the slope cannot be maintained, even ignoring the inertial loading on the soil due to the earthquake shaking, then large magnitudes of lateral spread displacement will be observed, as can be predicted from a Newmarkian sliding block analysis. (Newmark, 1965).

Fig. 4.1 Damage to a railway embankment during the 1964 Alaska earthquake due to lateral spreading of a river valley. (Photo from USGS).

Fig. 4.2 Damage to a railway bridge during the 1964 Alaska Earthquake (USGS).

The Newmarkian sliding block analysis models the slope as a rigid block of soil sliding on a rigid slope with the same inclination as the slope being analysed. Limiting accelerations, being the point at which the strength of the soil is just sufficient to allow the soil block to move in unison with the base slope, are calculated. If these limiting base accelerations are exceeded at the base of the soil mass, the difference between base acceleration and limiting acceleration is integrated to give an accumulated relative velocity, which then leads to accumulated downslope movement until the relative velocity returns to zero. This is shown schematically in Fig. 4.3.

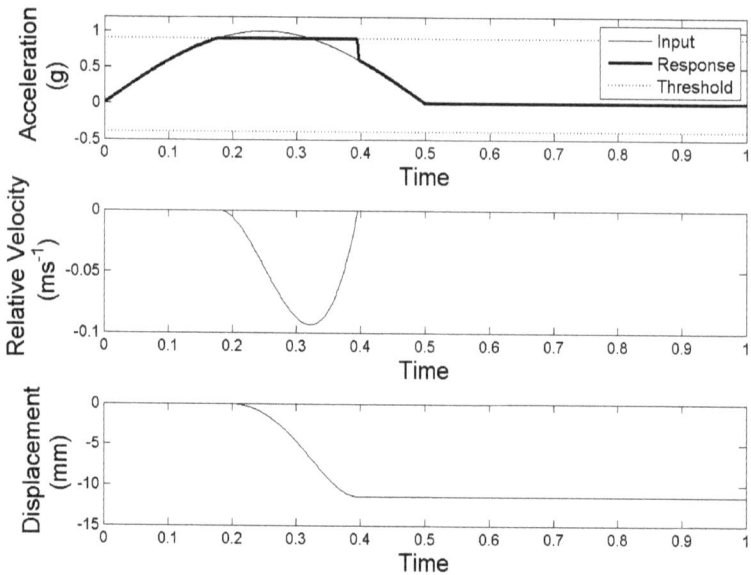

Fig. 4.3 Newmarkian sliding block mechanics.

During liquefaction, the threshold acceleration levels decrease due to increased pore pressures and loss of strength, resulting in increased observed deformation. This can be incorporated into a Newmarkian style analysis, as discussed in Haigh *et al.* (2001). If the pore pressure increase is sufficient that the threshold accelerations required to initiate downslope movement fall below zero, implying that the slope is unstable even without the disturbing effects of the earthquake, then lateral spread displacement will continue to accumulate until consolidation of the liquefied soil layer progresses sufficiently that enough strength is regained for static stability to be satisfied. In practice this implies that if approximately 5% of shear strength remains once the soil is fully liquefied, then slopes with angles greater than approximately 1.5° may suffer significant displacement due to lateral spreading.

Even if the residual strength of the soil is sufficient to maintain stability of the slope without the disturbing effects of the earthquake, some slope displacement may be observed when equilibrium cannot be maintained during the peak acceleration phases of the earthquake. This will result in a stick-slip motion of the slope, which may be very damaging to pile foundations, as the soil that is moving still retains significant strength and can hence exert significant forces onto pile foundations passing through it.

Observation of the pore pressures within laterally spreading slopes shows that the soil within the slope does not remain in a state of full liquefaction throughout the earthquake loading. The reversal of direction of shear stress that occurs due to the superposition of static shear stresses from the sloping ground and cyclic shear stresses from the earthquake loading can cause the soil to dilate locally, leading to a decrease in pore pressure and an increase in effective stress and strength. This 'de-liquefaction shock-wave', as described by Kutter and Balakrishnan (1999), can then propagate upwards through the soil leading to the slope regaining strength and downslope movement halting. This increase in strength while downslope movement is occurring can lead to large transient loads being applied to piles. This can be seen in the centrifuge test data shown in Fig. 4.4 in which the upward propagation of the negative pore-water pressure spike can clearly be seen.

Damage due to lateral spreading may also be seen in level ground if the stability of that level ground is provided by an earth-retaining structure such as a quay wall or retaining wall. Movement of the retaining structure may allow lateral movement of the retained liquefied soil, resulting in lateral spread displacements.

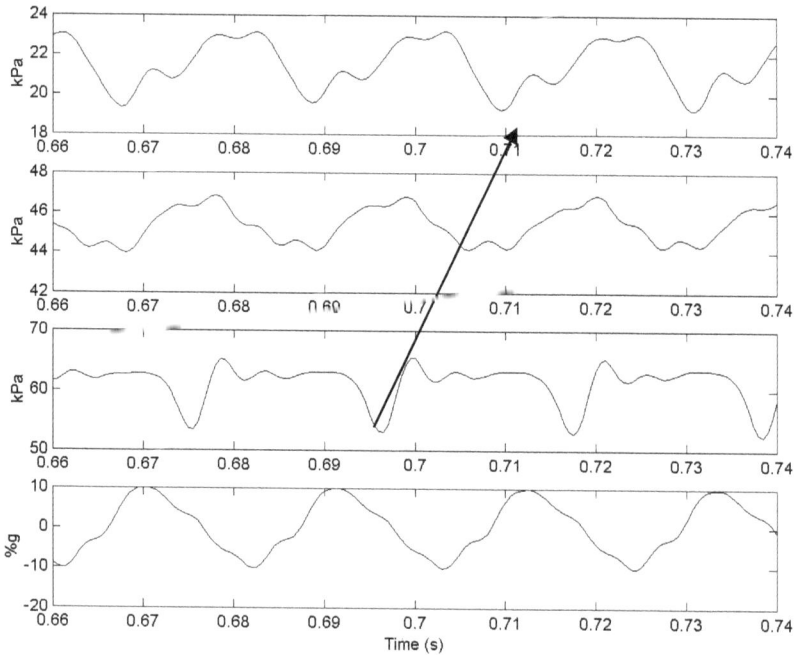

Fig. 4.4 Propagation of de-liquefaction shock wave upwards through a sand layer.

Calculation of the magnitude of lateral spreading may be achieved using this method if the base acceleration, residual strength and pore pressure history are either known or can be estimated. As this is often not the case, the simple empirical relationships presented in the next section may be useful in order to estimate the magnitude of lateral spreading displacements that might occur in a given situation.

4.2 Simple Methods to Estimate the Extent of Lateral Spreading

Several methods are available in the literature for the estimation of the magnitude of lateral spreading that may be suffered by sloping ground. Many of these were estimated from the analysis of lateral spreads from past earthquakes, with regression analysis being used to give a best fit to available data.

The simplest of these formulas is that derived by Hamada *et al.* (1986) based on lateral spreads observed from aerial photographs during the 1964 Niigata and 1983 Nihonkai-Chuba earthquakes. From this data the Eq. 4.1 was developed, in which D_H is the horizontal displacement at the ground surface in meters, T is the thickness of the liquefiable layer and θ is the gradient of the ground surface or of the base of the liquefiable layer, whichever is greater, in percent.

$$D_H = 0.75 \cdot T^{0.5}\theta^{0.3} \tag{4.1}$$

From this equation it can be seen that for a 5m thick liquefiable layer at a 6° slope angle (10% slope), lateral spreads of approximately 3.5m are predicted. As seen in the previous section, experimental data using centrifuge testing suggests that this may be conservative, as slopes of loose liquefiable sand using this geometry under severe shaking for approximately 30 seconds generally exhibit spreading of approximately 1.5m. This equation does not, however, involve any parameters due to ground shaking and as such may be of limited use in practice, as for areas with low seismicity the spreads predicted may be overly conservative.

Bartlett and Youd (1995) carried out a much larger exercise using multiple-regression analysis on data from 467 lateral spreads from both Japanese and US earthquakes. This data was used to generate Eq. 4.2.

$$\log(D_H) = -15.787 + 1.178M - 0.927\log(R) - 0.013R \\ + 0.348\log T_{15} + 4.527\log(100 - F_{15}) - 0.922D50_{15} \tag{4.2}$$

D_H is the horizontal ground displacement in meters, M is the earthquake moment magnitude, R is the horizontal distance to the nearest fault rupture in km, S is the ground slope in %, T_{15} is the thickness of saturated layers with $(N1)_{60}$ <15 in m, F_{15} is the average fines content in the liquefiable layer in percent and $D50_{15}$ is the average D_{50} grain size in the liquefiable layer in mm.

Thus for the same 5m thick liquefiable layer at a 6° angle (10% slope) if we assume a clean sand with D_{50} of 0.1mm and an SPT blow-count of 8 subjected to a magnitude 6.5 earthquake at a distance of 20 km, we predict a lateral displacement of 3.85m.

From the dataset used to generate the equation, the majority scatter of predicted versus measured displacement falls within a factor of two, as can be seen from Fig. 4.5. The equation thus gives a good estimate of the lateral spread that might be exhibited by a slope but the result given should be multiplied by two to give a more conservative design estimate.

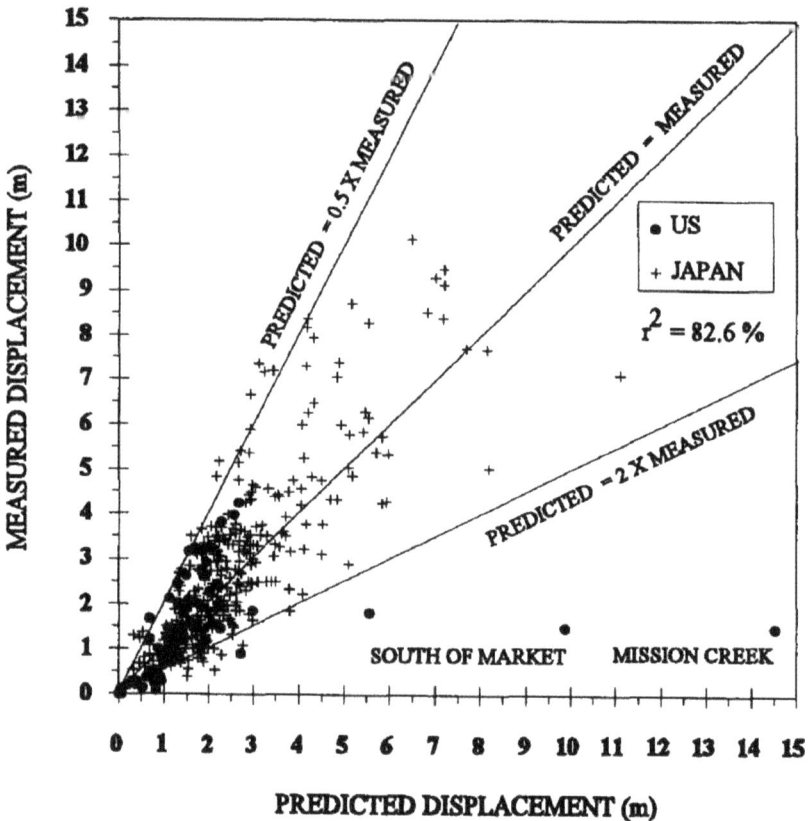

Fig. 4.5 Comparison of predicted versus measured displacements after Bartlett and Youd (1995).

Horne and Kramer (1998) derived a finite element programme called WAVE, which is a one-dimensional ground response model. The simplification to one spatial dimension is used to minimise the run time and storage space required and to avoid the complexity of determining a multidimensional constitutive model for the soil. A finite difference scheme was used to solve the equations of motion of the system using a constitutive model based on a two parameter hyperbolic function, with strain reversal hysteresis dealt with using the Cundall–Pyke hypothesis (Pyke 1979). According to this hypothesis, a strain reversal causes unloading on a scaled version of the original 'backbone' curve. Generation of excess pore pressure was dealt with using a modified version of the Nemat–Nasser and Shokooh (1979) energy scheme. This relates the work done during cyclic shearing to volume change and pore pressure change occurring during the shearing.

The predicted displacement depth relationship for a 20° slope is shown in Fig. 4.6, the multiple lines on the graph representing the variation of displacement and shear stress with time through one cycle of an earthquake.

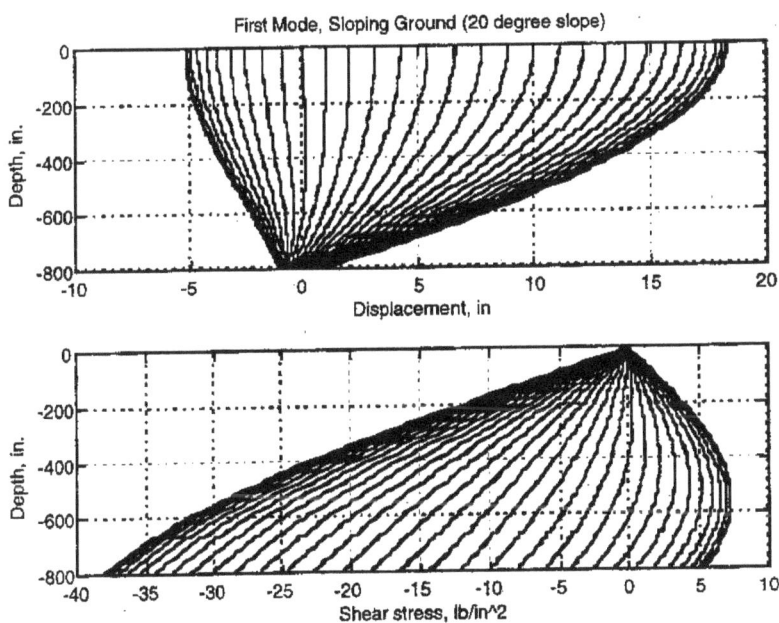

Fig. 4.6 Predicted spreading of a 20° slope. (Horne and Kramer, 1998).

Lateral spreading may also occur in level ground behind a waterfront retaining structure such as a quay wall. Tokimatsu and Asaka (1998) give equations linking the soil displacement behind the wall to the distance from the retaining wall and depth from the soil surface. This displacement is given by Eq. 4.3 and 4.4.

$$\frac{D(X)}{D_0} = 0.5^{5\frac{X}{L}} \tag{4.3}$$

$$f_{ls}(z) = D(x)\cos(\frac{\pi z}{2H}) \tag{4.4}$$

Where D(X) is the soil surface displacement a distance X behind a retaining wall whose displacement is D_0 with a liquefied layer thickness of H and the lateral spread propagating back from the wall by a distance L which is approximately 25–100 times D_0. $f_{ls}(z)$ gives the variation of displacement with depth as being a half cosine wave.

4.3 Effects of Lateral Spreading on Pile Foundations

Piles and pile groups passing through flowing layers of soil are obviously subjected to drag loading by the soil flowing around them. These loads are dependent on intrinsic soil parameters, soil effective stress state and on relative displacement that occurs between soil and pile. The magnitude of lateral loads that can be applied to piles are limited by the passive pressure that can be developed within the soil. The weak nature of the flowing liquefied soil (owing to its near-zero effective stress state) therefore limits the magnitude of the pseudo-static loading that can be applied to the piles by the flowing soil. It has been seen, however, that very high transient loads may be applied to the pile owing to local dilative behaviour as the soil is strained while passing around the pile, as will be discussed in Sec. 4.3.2.

The case history of the Showa bridge during the 1964 Niigata earthquake presented in Chapter 1 shows how the loads exerted by flowing liquefiable soil may be sufficient to cause significant damage to pile-founded structures. Figure 4.7 shows the movement of the banks of the Shinano River in Niigata due to lateral spreading, which occurred during the earthquake. Post-earthquake surveys showed that the river had narrowed by up to 11m. (Hamada *et al.*, 1992). This very large magnitude of soil movement obviously exerted large lateral forces on the piles of all the bridges crossing the river and resulted in the collapse of several of them.

Fig. 4.7 Ground movements in cm in Niigata during the 1964 Earthquake. (Hamada, 1992).

4.3.1 Presence of nonliquefiable crust

As discussed in the previous section, the lateral pressures applied by flowing liquefied soil are limited by the very low effective stresses existing in the liquefied soil. If a non-liquefiable crust exists on top of a liquefiable layer, by virtue either of the nature of the soil or of the depth of the water table, then significantly higher stresses may be exerted on the pile, as the stability of the slope is governed by the low strength of the liquefiable layer, whereas the loads acting on the pile are governed by the strength of the much stronger crust. This situation is the most critical for piles passing through laterally spreading deposits, as it is only in the presence of a non-liquefiable crust that very high lateral loads can be applied to the piles.

In these circumstances it would be rational to assume that the crustal soil exerts a lateral pressure equal to the full passive pressure on any pile or pile cap that passes through the non-liquefied crust.

4.3.2 Lateral pressures generated on piles and pile caps

In order to design pile foundations passing through laterally spreading liquefiable layers it is obviously necessary to quantify the loads that might be applied to those foundations by the flowing soil. The loading exerted on the foundation system can be divided into two categories:

- load exerted on the piles by the flowing ground; and
- load exerted on pile caps.

As mentioned in Sec. 4.3.1, the pressures exerted on foundation systems when a non-liquefiable crust overlays the liquefiable layer are significantly greater than those in the absence of this non-liquefiable crust. This is especially true when a pile cap is present, embedded into the soil. The pile cap will by necessity be embedded in the strong, non-liquefied crust and will present a large facial area on which the laterally spreading crust can exert pressure. The large lever arm at which this force is exerted relative to the base of the piles will result in enormous bending moments being applied to the piles.

This problem can be exacerbated by non-liquefiable surface soils, which are often present owing to either dense surface soils or to the water table being significantly below the soil surface. As the surface soil does not liquefy, it is able to apply large passive pressures to the pile cap as the soil beneath it liquefies and the un-liquefied crust moves downslope.

The problems caused by this situation were illustrated by the failure of the piles of the NFCH building during the Niigata earthquake. (Hamada *et al.*, 1986). Analysis of the foundation piles using the finite difference code B-STRUCT, which estimates the lateral spread pressure using *p-y* curves reproduced the damage exerted to the foundation piles by the 0.66m of lateral spread that was measured post-earthquake. (Meyersohn, 1994). The analysis involved taking an elastic-perfectly-plastic *p-y* curve defined for the non-liquefied soil (characterised by a limiting pressure p_u and a stiffness k_h, as shown in Fig. 4.8), and scaling down this curve by a reduction parameter R_f in order to account for the loss of strength in the soil due to liquefaction. It was found that to best reproduce the damage observed by post-earthquake excavation of the piles, a reduction factor of 60 should be used, implying that the soil had a strength of approximately 1.5% of that of the non-liquefied soil.

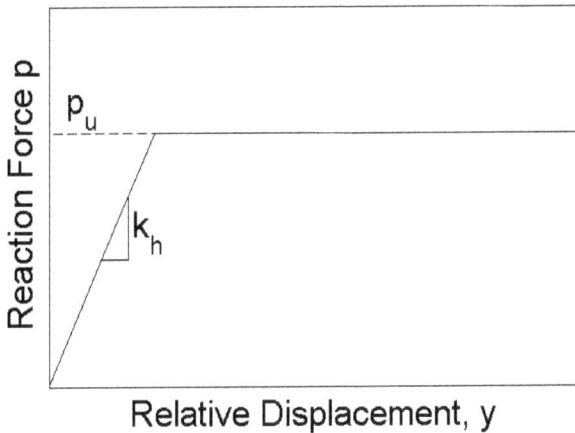

Fig. 4.8 *p-y* curve used in the B-STRUCT code.

Parametric studies yielded a diagram of failure modes, showing the transitions between bending, buckling and soil flow dependent on relative pile stiffness and axial load (Fig. 4.9). The responses of steel piles and pile groups were also investigated. When considering pile groups, the piles were assumed to be close together and to prevent the soil between them from liquefying. The pile group and the soil within it thus deforms as a single large-diameter object.

Increasing Soil Stiffness

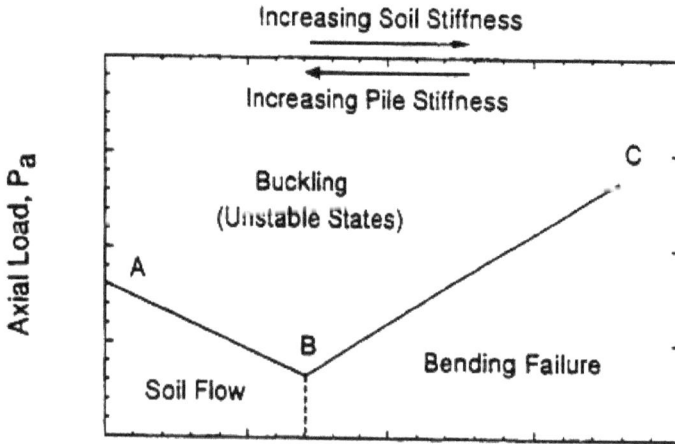

Fig. 4.9 Failure regimes of piles subjected to lateral Spread. (Meyersohn, 1994).

The lateral pressures that can be exerted onto pile foundations are obviously limited by the full passive pressure of the soil. If the soil were fully liquefied, the effective stress would be zero and hence the maximum lateral stress would be zero. Dilation of the soil around the pile, however, leads to a transient drop in excess pore pressure, a temporary increase in effective stress and hence horizontal pressures being applied to the piles. This phenomenon has been investigated using centrifuge modelling, including work at RPI by Abdoun (1997). The data obtained from this centrifuge model testing were used to generate recommendations as to the lateral stresses to be applied to piles. Recommendations of researchers at RPI have variously suggested that piles in laterally flowing soil be designed as being subject to constant pressures of 9.5kPa (Abdoun, 1997) or 11kPa (Ramos *et al.*, 1999) with depth through the liquefiable deposit or an inverse triangular distribution with depth falling from 17kPa at the surface to zero at 6m depth. (Dobry and Abdoun, 1998). The data in this study was derived by measurement of bending moments induced within model piles, with spline curves fitted to the bending moment distributions and used to calculate equivalent linear pressure distributions on the pile.

A numerical code based on p-y curves has also been developed by Goh and O'Rourke (1999) calibrated with the centrifuge results of Abdoun (1997). This code uses linearised versions of the strain-softening normalised p-y curves shown in Fig. 4.10.

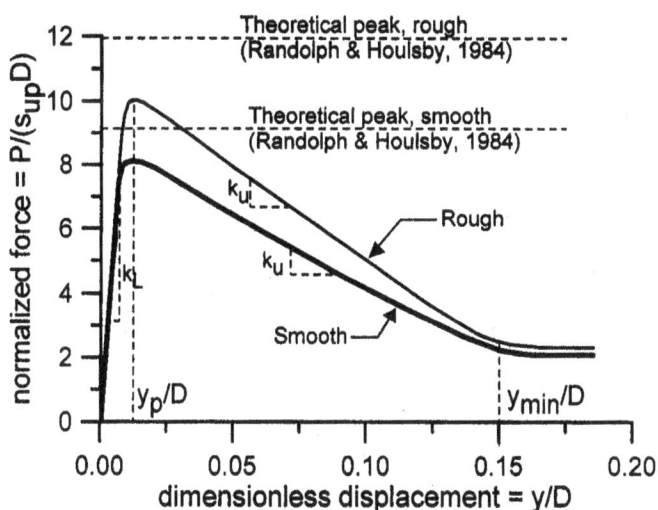

Fig. 4.10 Strain-softening p-y curves. (Goh and O'Rourke, 1999).

These are obtained from a FLAC finite difference code analysis of a pile being displaced through a cohesive material, with undrained shear strength falling linearly with plastic deviatoric strain until a residual value of shear strength is achieved as shown in Fig. 4.11.

This analysis gives a good fit to the centrifuge data with which it was calibrated, although it makes no attempt to predict any dynamic effects. The code was calibrated by changing the values of peak strength, residual strength and the displacements at which these are achieved and getting a best-fit to the centrifuge data. As there are four parameters to be chosen, a good fit is obviously achieved with the data with the best values of these parameters, regardless of the competence of the code.

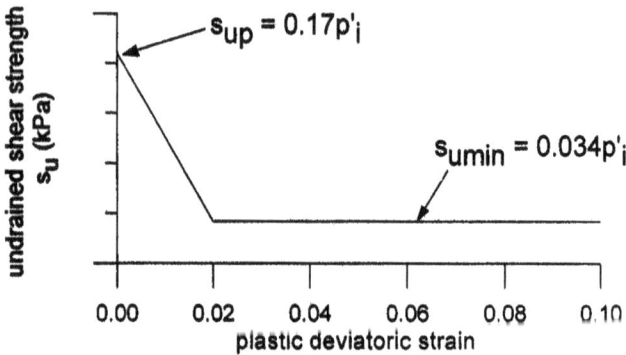

Fig. 4.11 Assumed soil stress-strain behaviour. (Goh and O'Rourke, 1999).

As the cyclic effects in the centrifuge data are approximately one third of the maximum bending moments measured, the authors conclude that pile response during lateral spread is controlled principally by the static load component. The analysis makes no attempt to model the true soil behaviour that occurs during shaking, assuming the softening of the 'springs' linking soil and pile to be only a function of relative soil-pile displacement. In reality, much of this softening is due to the liquefaction of the soil, a feature of earthquake loading, rather than relative displacement. It will also be seen later in this chapter that there are significant effects in terms of the near-pile soil effective stress state caused by dilative behaviour due to cyclic loading. These features are not considered by this analysis.

This work largely agrees with that by Liu and Dobry (1995), who used the results of a series of centrifuge model tests involving piles passing through liquefiable soils to back-calculate the reduction in soil strength with increasing pore pressure, as seen in Fig. 4.12. The coefficient c_u is the ratio of strength at a given pore pressure ratio to that with no excess pore pressure. It can be seen that when a value of r_u of 90% is achieved, the soil still maintains around 20% of its initial strength. Extrapolating this to a value of r_u of 100% leads to a residual strength of around 8% of the initial value, though this is not supported by any experimental data beyond $r_u = 90\%$.

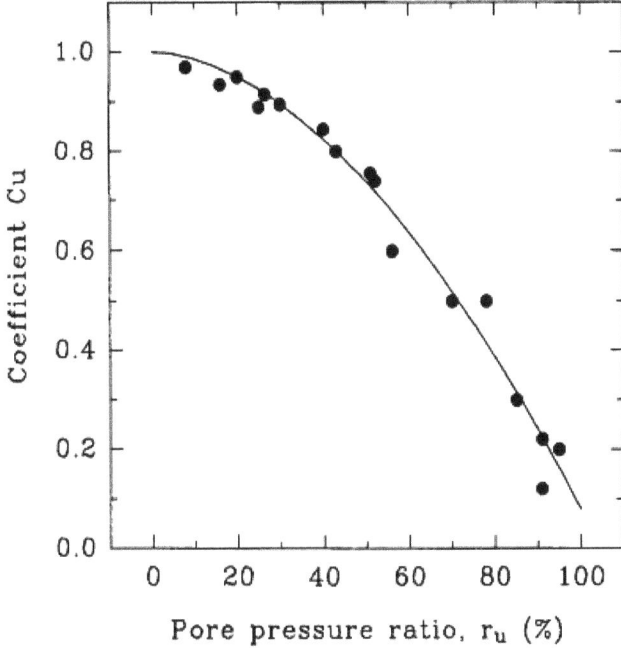

Fig. 4.12 Degradation of strength with excess pore pressure. (Liu and Dobry, 1995).

Centrifuge modelling tests at Cambridge University have built on this previous work by increasing the instrumentation in the soil close to the piles in order to study the soil behaviour more closely. A 6m-thick sand layer was constructed with a 6° slope, through which passed piles. Pile bending moments were measured, as in previous tests, but the soil pore pressures and lateral stresses close to the upslope and downslope faces of the pile were also monitored, as shown in Fig. 4.13. The net downslope pressure acting on the piles is shown in Fig. 4.14. It can be seen that while the residual pressure acting on the pile once shaking ceases is approximately 25kPa, significantly higher stresses than this (50kPa) can be measured transiently during the earthquake.

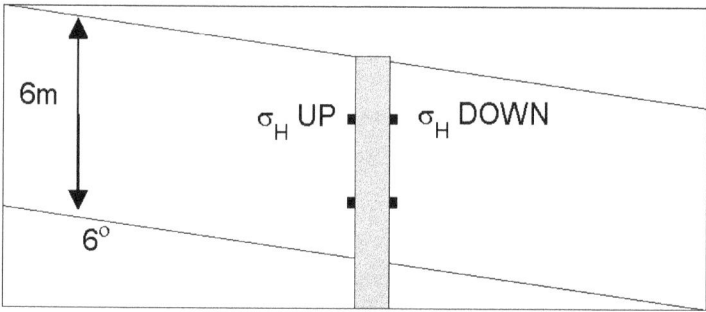

Fig. 4.13 Layout of centrifuge test model.

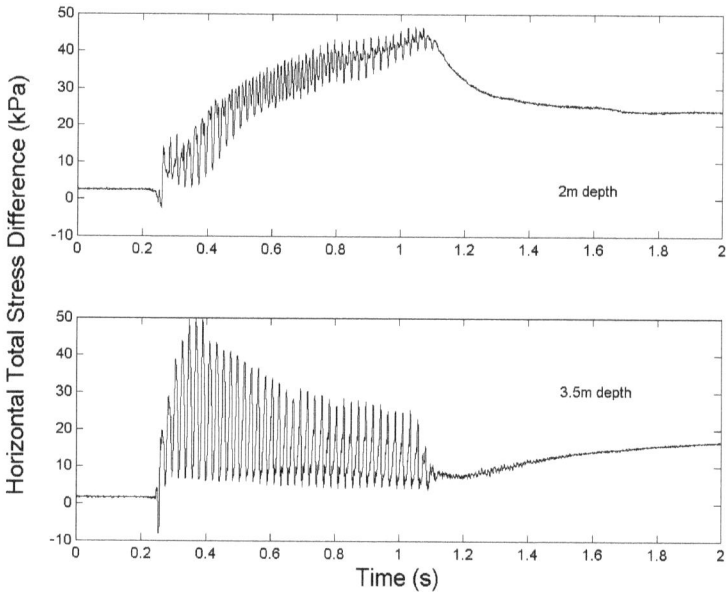

Fig. 4.14 Net downslope pressure acting on pile.

The bending moments measured within the piles (Fig. 4.15) however, do not show such a significant difference, peak bending moments exceeding residual ones by only about 10%, as the pile is not necessarily in equilibrium during the earthquake. In order to generate bending strains within the pile, it is necessary for the pile to move and deform. If insufficient time is available for this to occur, based on the pile's natural period, the pile may not be required to carry all of the forces exerted upon it, but can deform to limit damage. Back-calculation of the lateral pressures required to generate the bending moments observed in the piles revealed these to be approximately 25kPa, similar to the measured residual pressures.

The experiments also investigated the difference in bending moments exerted on piles with different cross-sectional shapes (circular and square, Fig. 4.15) and concluded that, while square piles offered more resistance to the flow of liquefied soil, this only resulted in an increase in bending moments of around 10%, similar to the magnitude of increase that could be predicted from plasticity-based solutions to the lateral flow problem.

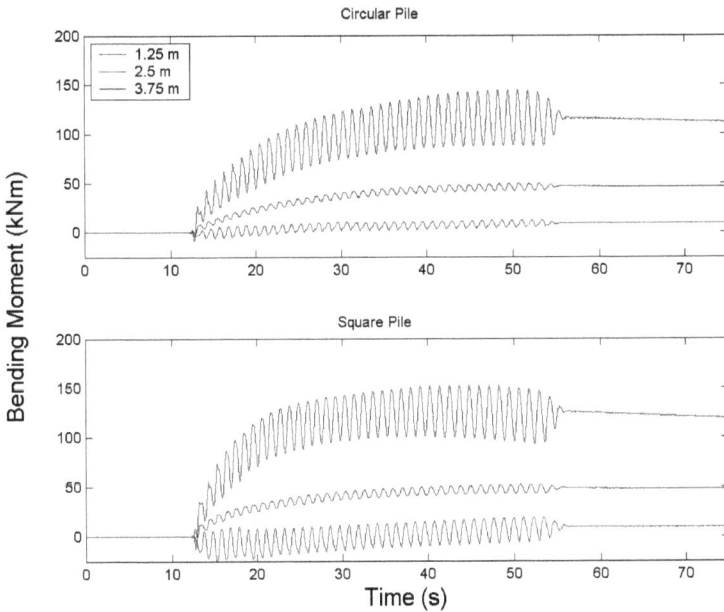

Fig. 4.15 Bending moments exerted on piles during centrifuge tests.

4.3.3 Current codal provisions

The design of piles passing through liquefied soil layers is codified in several Japanese design codes, including those for highway bridges, railway facilities and building foundations. These codes are discussed in detail by Uchida and Tokimatsu (2005). The differing approaches taken by these three codes are summarised below:

4.3.3.1 Specifications for Highway Bridges (JRA, 2002)

In laterally spreading ground a horizontal earth pressure is applied to the piles without inertial forces. This pressure is mandated to be:

- full passive earth pressure in the non-liquefiable crust
- 30% of total vertical overburden stress in the liquefied layer.

While liquefiable sands are frictional materials and would normally be assumed to apply pressures that are a function of effective, not total stress as at liquefaction, the effective stress in the liquefiable layer is zero. It is not sensible to define lateral pressures with respect to effective stress unless these are independent of depth.

4.3.3.2 Design Standard for Railway Facilities (RTRI 1999)

Piles should be designed using the seismic deformation method. The displacement due to lateral spreading should be calculated based on the thickness of the liquefiable layer and the predicted movements of any supporting retaining wall structures. The forces are calculated based on a subgrade reaction coefficient with a value one thousandth of that of the non-liquefied soil. Using the equation quoted by Uchida and Tokimatsu:

$$k_H = 3ND^{-3/4} \tag{4.3}$$

where k_h is the subgrade reaction coefficient, N is the SPT N value for the soil and D is the pile diameter in m. This implies that for a 30cm pile subjected to 1m of lateral spread in a soil with an SPT N-value of 10, a lateral pressure of 74kPa will act on the pile. This seems reasonable when compared with experimental data.

This design code will predict lateral pressures proportional to the lateral spreading displacement and so will tend to overestimate the lateral loads at high lateral spreads. Once the displacement exceeds a few pile diameters, it seems implausible that the load will continue to increase, but this will not be predicted by this design code.

4.3.3.3 Recommendations for Design of Building Foundations (AIJ 2001)

As in the RTRI code, design is based on a subgrade reaction coefficient. The coefficient can be found for a single pile from Eq. 4.4 (Uchida and Tokimatsu 2005):

$$k_H = 56000 N B^{-3/4} y^{-1/2} \tag{4.4}$$

where k_h is the subgrade reaction coefficient, N is the SPT N value for the soil, B is the pile diameter in cm and y is the soil displacement in cm.

This equation gives very large lateral pressures when lateral spreading occurs, which seems unrealistic given the low strength of the liquefied flowing soil.

4.3.4 Recent experimental data vs codal provisions

Recently, experiments have been carried out using large shaking tables by Suzuki *et al.* (2005) and centrifuge models by Haigh (2002) to investigate the behaviour of soil close to piles during lateral spreading. These experiments have measured lateral stresses exerted on piles, bending moments in piles and pore pressures in the soil close to both the front and back faces of the piles.

The experimental evidence shows that during shaking there is a great deal of dilation of the soil close to the pile, especially on the downslope side. This localised dilation results in the soil close to the pile retaining significant strength throughout the earthquake loading, resulting in enhanced loading on the piles. The experiments also showed that the dynamic behaviour of piles plays a significant role in the transfer of lateral loads from the soil into pile bending moments.

During earthquake loading, the pile foundation is not necessarily in equilibrium, as it has its own acceleration. This may result in the bending moments in the pile being significantly less than those which would be found from integration of the lateral loads placed on the pile. It is the bending moments that cause damage to the pile and hence this flexibility may protect the piles from some of the peak dynamic and inertial loads that occur during shaking, though the large magnitudes of loading that may be experienced during post-earthquake flow will not be attenuated.

Measurements of the peak transient lateral stresses imparted on piles by the flowing liquefiable soil have been shown to be significantly larger than those predicted by design codes such as the Japanese Highway Bridges Code (JRA), but the transient nature of these loads may allow them to be sustained by the piles without significant damage. The pseudo-static loads exerted on the piles by the flowing soil agree much more closely with the codal provisions.

4.4 Recommendations on Estimation of Lateral Loads for Pile Design

This chapter has demonstrated several methods for predicting both the lateral spread displacements of liquefied ground and the loads that spreading soils exert on pile foundations passing through them. As the lateral spreading load will reach its maximum value at a relatively small displacement compared to the large magnitudes of lateral spreads that can be observed in the field, it seems more logical to design based on a limiting lateral earth pressure, rather than within a subgrade reaction-based framework. This has the double benefit of making the design of piles possible even when they are subjected to very large magnitudes of lateral spread and of removing the need to calculate accurately what lateral spread displacement will occur; it is sufficient merely to say that the slope could experience significant lateral spreading.

Based on the experimental data discussed in previous sections, it is recommended that the lateral loads exerted on piles by flowing liquefied soil be calculated based on full passive pressure within any non-liquefiable crust present above the laterally spreading liquefied layer and a lateral pressure of 20–30kPa in the liquefied layer. If low amounts of lateral spreading occur, the full lateral loads may not be exerted on the piles, as insufficient strain will be mobilised to apply the full pressure, however, this should be a conservative design approach.

From the data of a variety of experiments at both full-scale and in reduced-scale centrifuge models, this appears to be a conservative value for the lateral loads induced on pile foundations passing through these layers. Examples of design using this recommendation together with those from previous chapters will be given in Chapter 6.

Chapter 5

Axial Loading on Piles in Laterally Spreading Ground

5.1 Introduction

Chapter 4 has outlined the kinematic loads that laterally spreading soils impart on piled foundations and current methods for evaluating these. This chapter will begin by introducing some of the other loads that act on piled foundations in liquefiable soils during seismic events, including inertial forces and axial loads generated in the piles due to the presence of the superstructure. The relative magnitudes of these different loads at different times during earthquake shaking (load phasing) will then be discussed. Pseudo-static methods for evaluating foundation response to both peak transient and residual loads are presented and the effects of soil profile, foundation configuration and axial load are discussed.

5.2 Phasing of Loads

In Chapter 4, the kinematic loads generated on piled foundations due to soil-pile interaction (SSI) were discussed. In addition to these loads, the superstructure supported by the piles will also load the piles through their heads. These loadings may be divided into two main elements:

1. inertial loads; and
2. axial pile loads.

These will be discussed in the following sections in relation to the kinematic loads previously described in Chapter 4.

5.2.1 Inertial and kinematic loads

Inertial loads in piles react to the dynamic shear forces generated in the columns (or other vertical structural elements) due to dynamic deformation of the structure relative to the foundation. These loads are transient and exist only during strong shaking. In addition to the inertial forces generated due to vibration of the pile cap, these forces generate dynamic bending moments and shear forces within the piles. These forces are superimposed on the moments and shear forces generated by the relative soil-pile movement (kinematic loading, c.f. Chapter 4), which will have both a monotonic and dynamic component.

Comparing the inertial and kinematic components, the former tends to be high early in the earthquake, before the soil has generated significant excess pore pressures. This is because, prior to softening, the soil is relatively stiff and is effective in transmitting the ground motion into the structure. As the soil liquefies, the foundation undergoes larger relative soil pile movement due to the increasingly different stiffnesses of the soil and the piles. This tends to isolate the structure, so that inertial loads later in the earthquake are much smaller than those at the beginning. In direct contrast, dynamic kinematic loads increase during liquefaction as the soil becomes more flexible relative to the pile. If the soil profile is not level, spreading may occur, which will increase monotonically with increasing excess pore pressure and with earthquake duration when the soil is fully liquefied. This spreading component is superimposed on the dynamic kinematic loads. These effects are shown schematically in Fig. 5.1.

To illustrate this effect, centrifuge tests have been conducted by Chang *et al.* (2005) on 3×2 pile groups supporting superstructures with natural periods of $T_{n,s} = 0.3$ and 0.8 seconds. In these tests, the demand on the piles is expressed as a total shear force at the pile head (V) due to the inertial load (I) and the kinematic load (F_{crust}) from the surrounding spreading non-liquefiable crust. In a simple model for the maximum shear force applied to the pile during an earthquake (V_{max}), V_{max} may be related to the peak inertia force (I_{max}), and peak crust load ($F_{crust,max}$) by

$$V_{max} = \chi\left(I_{max} + F_{crust,max}\right). \tag{5.1}$$

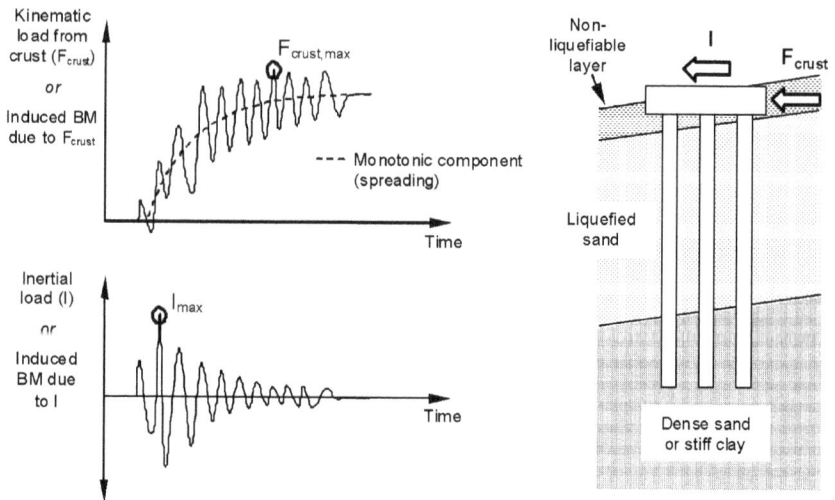

Fig. 5.1 Phasing of inertial and kinematic loads acting on piled foundations during liquefaction-induced lateral spreading.

Figure 5.2 shows the back-calculated values of χ measured by Chang *et al.* (2005) for two different time histories representative of the 1995 Hyogoken Nambu (Kobe) Earthquake and the 1989 Loma Prieta Earthquake. It can be seen from Fig. 5.2 that the peak shear force acting on the piles (i.e. the peak pile demand) is well approximated by adding the peak demand due to inertial and kinematic loadings separately (i.e. χ = 1). The mean value of the data presented in Fig. 5.2 gives χ = 0.94. These results agree with the findings of Tokimatsu *et al.* (2004) and Tamura and Tokimatsu (2005) for pile groups in level liquefiable ground. Tokimatsu *et al.* demonstrated that the peak inertial and peak kinematic forces should be added (i.e. χ = 1) to approximate the loading at peak pile demand when the natural period of the ground is longer than that of the structure. As soil liquefaction is accompanied by a reduction in stiffness and a consequent lengthening of the soil's natural period, the natural period of the (liquefied) spreading ground is expected to be larger than that of the superstructure in most cases, and supports the findings of Chang *et al.* (2005). Methods for estimating the peak dynamic kinematic and inertial load components are discussed in Chapter 2.

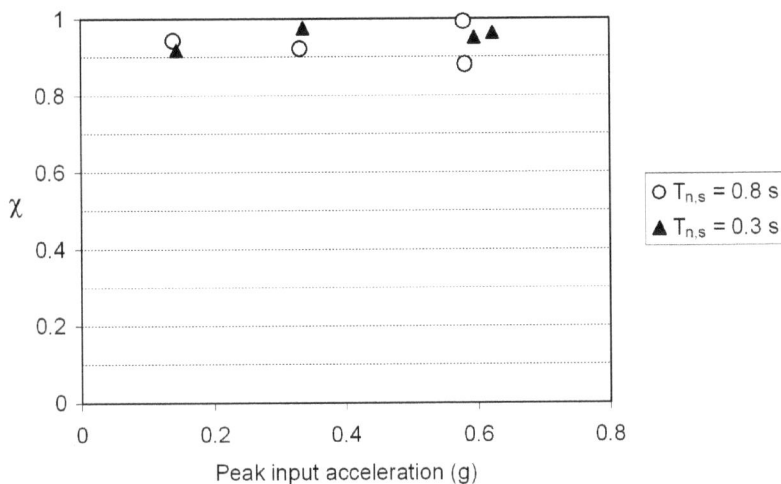

Fig. 5.2 Measurements of factor χ describing inertial and kinematic load superposition. (Data re-plotted from Chang *et al.*, 2005).

The foregoing discussion in this section has concentrated on peak foundation response to the maximum transient seismic loads occurring during earthquake shaking. Once strong shaking has ceased, there will be a residual foundation displacement due to the monotonic spreading component of the kinematic load (c.f. Fig. 5.1). Furthermore, the excess pore pressures within the soil will not dissipate immediately at the end of shaking due to the finite permeability of the liquefied soil. As such, pore pressures will remain high for a period of typically a few hours following the end of shaking. This will generally lead to further spreading of the soil and increased monotonic displacement of the foundation, which may be a more critical condition than the peak response during the earthquake. As a result, Tokimatsu and Asaka (1998) have recommended that two conditions are evaluated, namely:

1. peak response during shaking (inertial and kinematic loads); and

2. residual response at maximum spreading displacement (monotonic kinematic component only).

These two criteria are summarised schematically in Fig. 5.3.

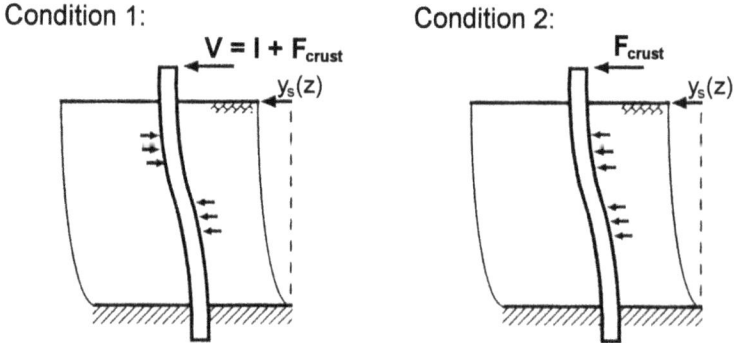

Fig. 5.3 Critical loading cases for piled foundations in laterally spreading soil.

5.2.2 *Presence of axial loads*

In order for inertial loads to be present, the supported superstructure must have some mass, which will also create axial loads in the piles. The axial load in a given pile will be composed of a monotonic and a cyclic component. The monotonic component consists of the initial static dead load, which is present in the pile irrespective of the earthquake, and a component due to gross rotation of the pile cap under the action of the monotonic component of the kinematic spreading loads. This latter effect occurs as the upslope piles are pulled upwards, reducing the compressive loads they carry, while those on the downslope side are pushed in, increasing the vertical load carried above the initial dead load.

Pamuk *et al.* (2003) have observed such behaviour in centrifuge tests of pile groups in laterally spreading soil. This is shown schematically in Fig. 5.4 for two piles carrying the same initial dead load. The cyclic component is induced due to the cyclic rocking of the pile group under the combined action of the inertial forces and the cyclic component of the kinematic forces.

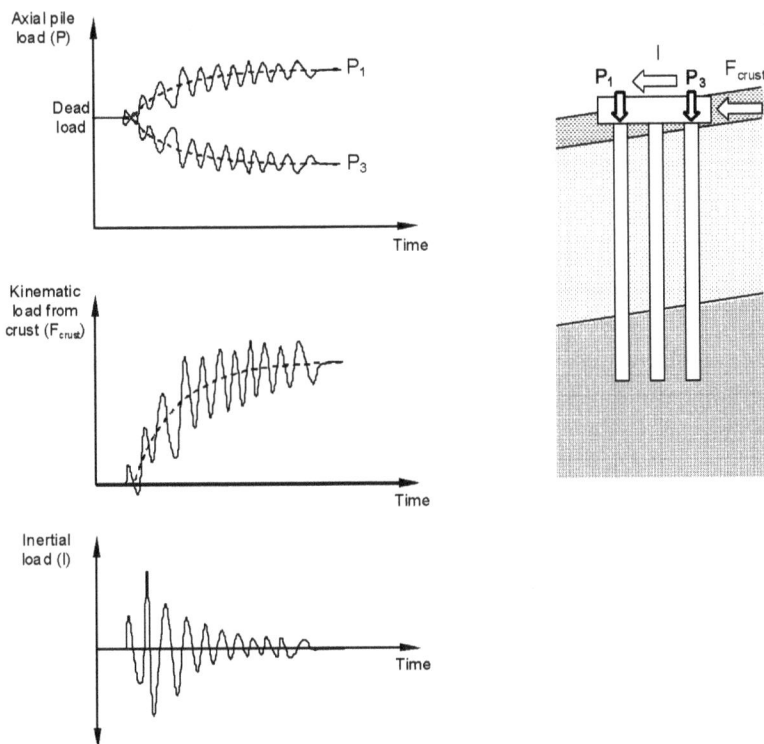

Fig. 5.4 Seismic axial loading of piles during liquefaction-induced lateral spreading.

5.3 Peak Lateral Response of Piled Foundations

At the end of Sec. 5.2.1, two limiting conditions were suggested which may lead to failure of piles, namely peak (co-seismic) lateral response and residual lateral response. This section will address the first of these conditions.

For important or unusual structures, the use of fully dynamic numerical (Finite Element) models or dynamic centrifuge testing is generally recommended. This can evaluate the full dynamic response from which the conditions at either maximum foundation displacement or maximum pile bending moments may be extracted. Such methods, however, are computationally and chronologically expensive and require skilled modellers in order to develop reasonable results. As a check on such methods however, and for the case of more standardised foundations, simpler pseudo-static methods can be used to predict the peak lateral foundation response. This makes use of the work of Chang *et al.* (2005), as outlined in Sec. 5.2.1, in that peak pile response is found by determining the response to the pile-head shear force V_{max}, Eq. 5.1, which incorporates the effects of both inertial and transient kinematic loads.

Pseudo-static 'push-over' methods assume that at any instant during an earthquake the piles are instantaneously in equilibrium. The equation governing the response of the pile is given by Eq. 5.2:

$$EI\frac{d^4y_p}{dz^4} + P\frac{d^2y_p}{dz^2} + k_l D_0 \left(y_p - y_s \right) = 0 \,. \tag{5.2}$$

where EI is the bending stiffness of the pile, $y_p(z)$ is the deflected shape of the pile, z is the depth below the ground surface, P is the axial load in the piles, $y_s(z)$ is the displacement of the surrounding soil, k_l is the soil-pile stiffness (stiffness of the *p-y* spring) in the liquefiable soil and D_0 the pile diameter. Equation 5.2 is a fourth order ordinary differential equation and as such requires four boundary conditions for its solution. Three of these may be selected from Table 5.1 (two at the tip and one at the top of the pile). The remaining boundary condition accounts for the inertial and crustal forces applied at the head of the pile, and is represented by

$$EI\frac{d^3y_p}{dz^3}\Bigg|_{z=0} = V_{max} \,. \tag{5.3}$$

where V_{max} is given by Eq. 5.1 (Sec. 5.2.1).

Table 5.1 Boundary conditions for piles for use in Eq. 5.2.

Location	Fixity condition	Boundary condition(s)	
Tip	Fixed	$y_p\big|_{z=L_p} = 0$	No displacement
		$\dfrac{dy_p}{dz}\bigg|_{z=L_p} = 0$	No rotation
	Pinned	$y_p\big|_{z=L_p} = 0$	No displacement
		$EI\dfrac{d^2 y_p}{dz^2}\bigg|_{z=L_p} = 0$	No bending moment
Top (sway permitted)	Fixed	$\dfrac{dy_p}{dz}\bigg|_{z=0} = 0$	No rotation
	Pinned	$EI\dfrac{d^2 y_p}{dz^2}\bigg|_{z=0} = 0$	No bending moment

The use of Eq. 5.2 assumes that the inertia of the pile itself is negligible compared with that imparted at the pile head (I_{max}). The second term, relating to the axial load P, is also usually ignored. As was explained in Sec. 5.2.2 however, the dynamic axial loads in the piles will generally be in phase with the inertial forces, so that it is prudent to also use $P = P_{max}$ when computing the response. It was shown in Chapter 3 (Sec. 3.4.1) that the peak cyclic load can be very much larger than the initial axial dead loads in the pile so that neglecting the second term may lead to inaccurate predictions of peak response. P_{max} may be estimated by considering the instantaneous rotational equilibrium of the pile cap under the overturning moment imparted due to superstructural inertial forces.

For simple soil profiles (e.g. pile surrounded by homogeneous liquefied soil), simple mathematical expressions for $y_p(z)$ and $y_s(z)$ may be used, so that Eq. 5.2 may be solved analytically. Suitable approximate deflected pile shapes, which satisfy fixity conditions at the top and tip of a pile are given in Table 5.2. Towhata *et al.* (1999) suggest that for a uniform profile of laterally spreading soil

$$y_s(z) = y_s\big|_{z=0} \, \sin\frac{\pi(L-z)}{2L}.$$ (5.4)

where L is the depth of the liquefiable layer. The value of y_s at $z = 0$ may be estimated using empirical methods such as those of Bartlett and Youd (1995), Rauch and Martin (2000) and Youd *et al.* (2002).

Table 5.2 Analytical approximations to lateral pile displacement, $y_p(z)$.

Fixity at tip	Fixity at top	Displaced shape
Fixed	Fixed (with sway)	$y_p(z) = \dfrac{y_p\big|_{z=0}}{2}\cdot\left[1+\cos\dfrac{\pi z}{L}\right]$
Fixed	Pinned (with sway)	$y_p(z) = y_p\big|_{z=0}\cdot\left[1-\sin\dfrac{\pi z}{2L}\right]$
Pinned	Fixed (with sway)	$y_p(z) = y_p\big|_{z=0}\cdot\left[\cos\dfrac{\pi z}{2L}\right]$
Pinned	Pinned (with sway)	$y_p(z) = y_p\big|_{z=0}\cdot\left[1-\dfrac{z}{L}\right]$

It is more common, however, for Eq. 5.2 to be solved computationally using finite difference methods, which may then account for variable soil properties/stratigraphy. This is more useful in situations where detailed soil data are available (e.g. from CPT tests), and is also able to account for discontinuities in displacement between liquefied soil and a non-liquefiable crust when water films are formed. By breaking down Eq. 5.2 in this way, it is additionally possible to determine the soil displacement profile $y_s(z)$ using numerical tools for determining the free-field site response. Such tools can also incorporate the effects of complex soil stratigraphy. Alternatively, the macro element models for pile head response and the effective length concept, both detailed in Chapter 2, may be used to determine the peak pile response and estimate the internal forces and bending moments within the piles.

The methods outlined in this chapter may also be used to determine the residual lateral response of the pile to spreading. This is accomplished by revising $y_s(z)$ to account for the maximum soil displacement that has occurred by the time the excess pore pressures have fully dissipated, and by setting $V_{max} = 0$ in Eq. 5.3. However, for some common soil profiles found in the field, simpler expressions have been found to predict the residual response, which can be solved more rapidly and more easily in closed form. These methods will be discussed in the following section and are particularly useful as they aid the understanding of some of the phenomena that can occur.

5.4 Residual Lateral Response of Piled Foundations

5.4.1 Single piles

Before investigating the response of pile groups including the effects of axial load, it is instructive to investigate the case of a single pile within a three-layer soil profile which is commonly found in the field. For the single pile, the analysis technique that is used here is that given in Sec. 2.8 . (Dobry *et al.*, 2003).

In Fig. 5.5, two different piles are considered to represent a flexible ($EI = 20$ MNm^2) and a stiff ($EI = 500$ MNm^2) pile. The other parameters were kept constant at: $L - 14$m, $h - 2$m, $p_0 - 150$kN/m^2 and $k_r = \infty$. The ultimate values $M_{a,ult}$ and $M_{b,ult}$ at $z_r = 0$ are 4400kNm and -200kNm respectively and are independent of EI. It can be seen that for the flexible pile, the bending moments are relatively low, so that $M_{a,ult}$ and $M_{b,ult}$ are reached at very large lateral displacement (not shown). In Fig. 5.5b the piles are much stiffer and attract greater bending moments for a given displacement. As a result, $M_{a,ult}$ and M_{bult} are reached at $\Delta \approx$

0.55m. However, it should be remembered that for a given applied spreading force from the soil, the deflection of the stiffer pile will be much lower.

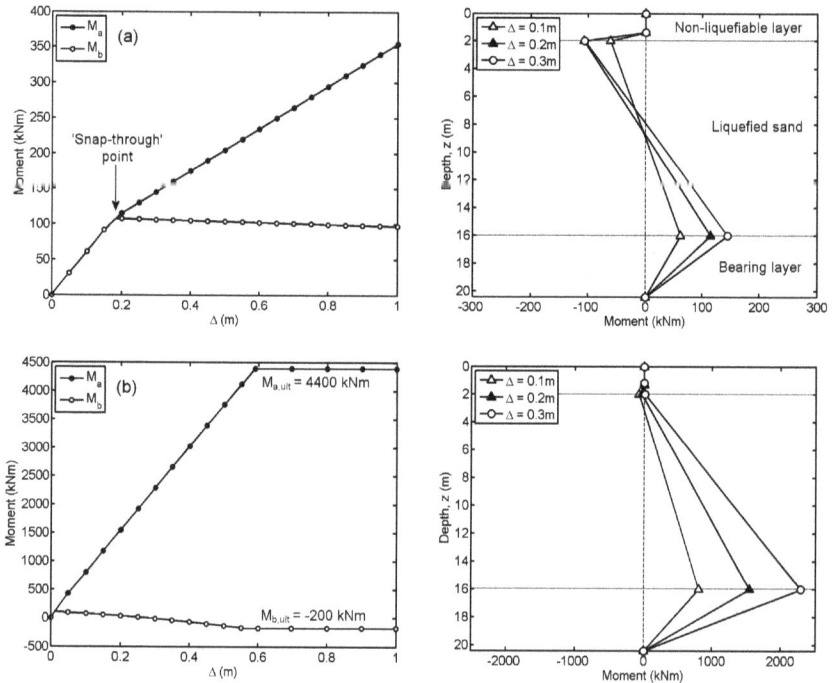

Fig. 5.5 Bending moments generated in a single pile with (a) $EI = 20$ MNm^2 and (b) $EI = 500$ MNm^2.

It can also be observed from Fig. 5.5 that the bending moments in the upper part of the pile are limited due to 'snap-through' of the pile through the clay layer (i.e. failure of the surrounding soil), so that the maximum moments occur at the interface between the liquefiable and bearing layers. This pattern of damage will be compared to that for piles in a group connected by a rigid pile cap in Sec. 5.4.3.

5.4.2 Pile groups (including axial load)

For piles in a group, connected by a pile cap, the fixity at the pile-cap interface dominates the behaviour of the pile towards the head. Under these conditions the pile cap translates laterally with little rotation and the 'snap-through' behaviour of the pile modelled by Dobry *et al.* (2003) is suppressed by the pile-cap fixity. The resulting displaced shape of the piles is as shown schematically in Fig. 5.6. As a result, it becomes possible to consider a simpler pile model in which the pile acts as a beam with full fixity at either end, but with relative sway between the two ends permitted. The fixity at the top of each pile is representative of the pile-cap connection mentioned above, and that at the bottom is indicative of the fixity depth of the pile within the underlying bearing layer. It is also possible with this model to consider the axial load carried by the pile.

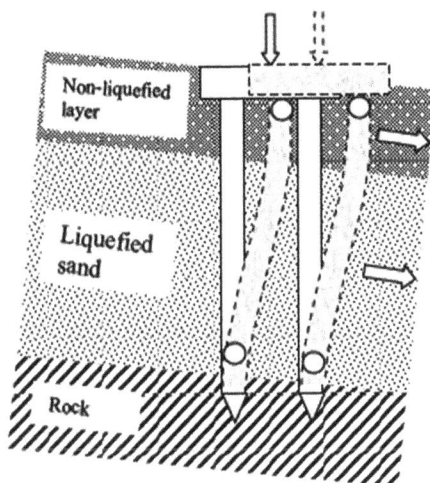

Fig. 5.6 Pile group failure in laterally spreading ground with non-liquefied crust (c.f. Sec. 1.6.2).

Limiting equilibrium (LE) may be used to pseudo-statically determine the response of the pile group to kinematic loads from the laterally spreading liquefiable soil and non-liquefiable crust. The model is shown schematically in Fig. 5.7. Soil pressures on the piles from the liquefiable layer (p_l) are considered to be constant with depth. The piles are considered to be flexible with a bending stiffness of EI and outer diameter D_0 and the pile cap is assumed to be rigid. It should also be noted that the model is based on purely lateral sway of the pile group (i.e. additional lateral displacement generated by rocking of the pile cap due to the axial compliance of the soil in the bearing layer is neglected).

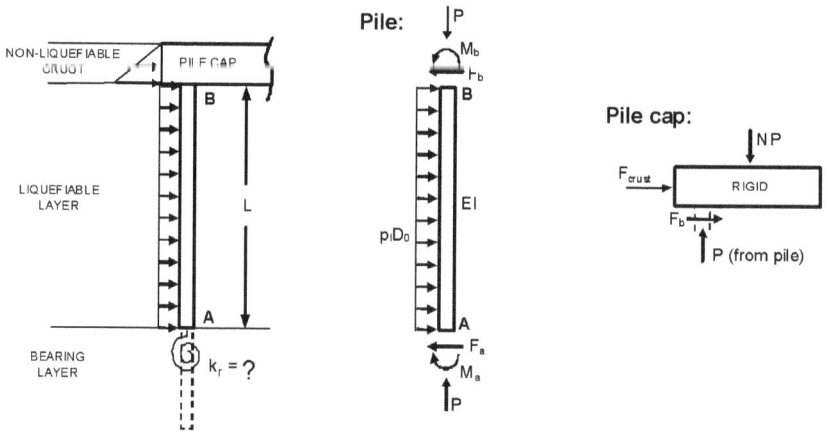

Fig. 5.7 Simple limit equilibrium (LE) model for estimating pile group response in laterally spreading soil, accounting for axial loads.

The equations governing the internal equilibrium of the pile may be expressed as:

$$M_a = -P\Delta \cdot f_\Delta - \frac{p_l D_0}{\mu^2} \cdot f_{pl} \qquad (5.5)$$

$$M_b = +P\Delta \cdot f_\Delta - \frac{p_l D_0}{\mu^2} \cdot f_{pl} \qquad (5.6)$$

$$F_a = \frac{M_b - M_a + p_l D_0 \frac{L^2}{2} - P\Delta}{L_{eq}} \tag{5.7}$$

$$F_b = -F_a + p_l D_0 L \tag{5.8}$$

where $\mu = \sqrt{(P/EI)}$, and f_Δ and f_{pl} are non-dimensional factors which may be found from Fig. 5.8. L is the length of the pile from the top of the liquefied layer to the fixity depth within the bearing layer.

If the liquefiable sand is expected to move further than the pile, p_l is taken to be positive (i.e. forcing the piles downslope in addition to the force from the non-liquefiable crust). In the case of relatively impermeable crustal layers (e.g. clays and silts), the upward hydraulic gradient generated as a result of liquefaction of the underlying layer leads to the development of a water film between the two layers. (Kokusho, 1999; Malvick *et al.*, 2006). In this case, lateral displacement of the crust can be very much larger than that of the spreading sand. This can force the pile cap (and hence the piles) to displace more than the liquefied sand, which can be modelled by using a negative value of p_l (i.e. acting upslope).

By considering the global equilibrium of the system, it is clear that the sum of the applied horizontal forces from the soil and structure is equal to the sum of the shear forces at the head of each pile (F_b). As inertial loads are usually transient with no residual component, the monotonic response of the piled foundation is due to the kinematic force from the non-liquefied crust alone, so that:

$$F_{crust} = \sum_{n=1}^{N} -F_{b,n} \tag{5.9}$$

where N is the total number of piles in the group.

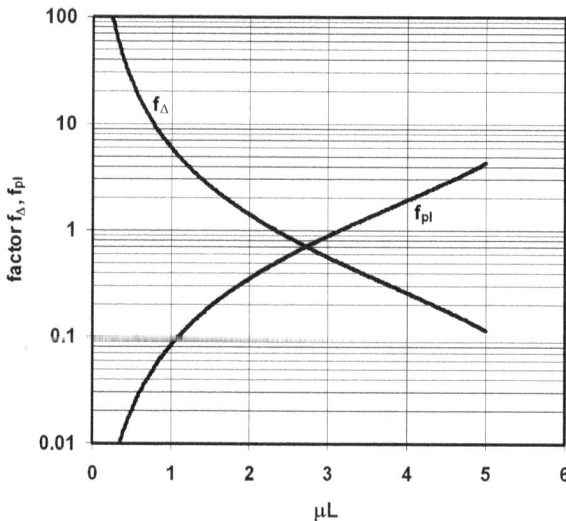

Fig. 5.8 Factors f_Δ and fpl for the analysis of pile groups accounting for axial load effects.

It is clear from Eqs. 5.5–5.9 that increasing the axial load in the piles increases the bending moments and reduces the applied crust load F_{crust} for a given amount of deflection. This means that as the axial load is increased, the piles will be less stiff laterally (i.e. for a given crust load the deflection will be greater).

The model as presented above is a fully elastic method. However, in reality the crustal load which can be applied to a pile group will be limited by one of two factors, namely:

1. limiting moment/shear force (structural capacity) of the piles; or
2. yield in the surrounding crustal soil (occurring at $F_{s,ult}$).

The first of these effects will lead to large pile group deflections when the applied F_{crust} is sufficient to cause failure of the piles. The final deflection of the pile group in this case will depend on the amount of subsequent motion of the crustal layer (the soil layer and the pile will move together after pile yield). For the second effect, when F_{crust} to cause soil yield is lower than that to cause pile yield, foundation deflections will be significantly smaller. These two effects are shown schematically in Fig. 5.9.

The value of F_{crust} may be easily evaluated from Eqs. 5.5–5.9 within a spreadsheet for a range of values of Δ to determine the overall pile group lateral load-deflection response. In doing this, values of the bending moment at each end of the pile M_a and M_b will be computed using Eqs. 5.5 and 5.6. At each step in the analysis, failure of the pile may be checked either by comparing the moment distribution within the piles to the yield moment (ductile piles), or by comparing the shear force distribution within the piles to the ultimate shear force (brittle piles failing in shear). Similarly, failure of the soil may be ascertained by comparison of F_{crust} to the yield force in the crustal soil ($F_{s,ult}$). The soil yield force $F_{s,ult}$ may be found using the methods presented in Chapter 4. By including these two effects, the model becomes elasto-plastic and is able to replicate the behaviour of foundations in laterally spreading soil from centrifuge tests (see Sec. 5.5). Compatible crustal soil displacements (shown schematically in Fig. 5.9) assuming elastic behaviour prior to yield may be found for given values of F_{crust} and Δ (structural response) computed using Eqs. 5.5–5.9, as

$$F_{crust} = k_{crust}\left(y_{crust} - \Delta\right) \tag{5.10}$$

where k_{crust} is the elastic soil stiffness (dependent on the shape of the pile cap) and y_{crust} is the lateral displacement of the spreading non-liquefiable crust. From the form of Eq. 5.10 it can be seen that the deflection of the surrounding non-liquefiable crust will always be larger than the pile-head deflection while the pile responds elastically (shown schematically in Fig. 5.9).

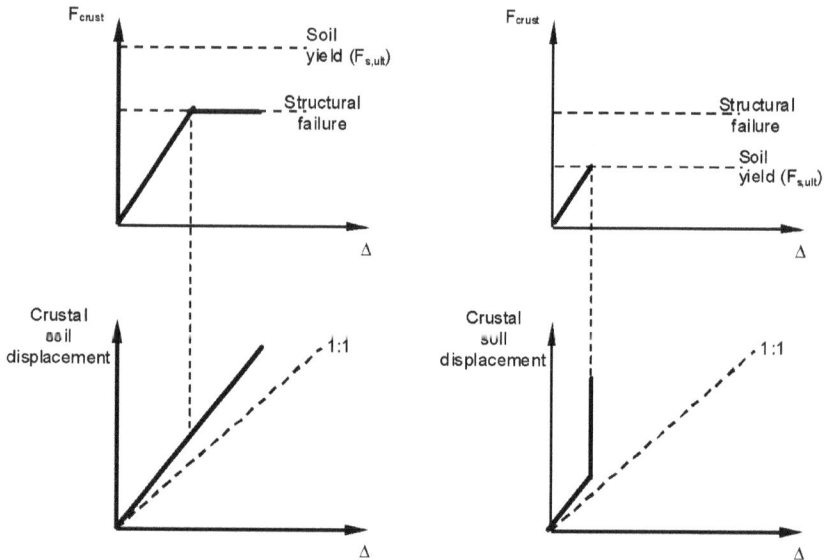

Fig. 5.9 Inclusion of inelastic effects.

5.4.3 Comparison of single pile and group pile behaviour

A comparison of the LE models for both the single pile and for a similar pile in a pile group is presented in Fig. 5.10. The parameters were kept the same as the example in Sec. 5.3.1 (EI = 20 MNm^2, L = 14 m, h = 2 m and p_0 = 150 kN/m^2, p_l = 0). In the case of the pile group, the pile cap was assumed to be the same thickness as the crustal layer (i.e. 2m).

The ultimate capacity of the pile was M_{ult} = 160kNm. Failure of the piles will occur when the ultimate capacity is exceeded. This occurs at the bottom of the liquefied layer in the case of the single pile, while the moments at the interface between the crust and liquefied sand are capped due to 'snap-through' of the pile. In contrast, for the same amount of lateral deflection, failure of the pile connected to the pile cap occurs simultaneously at both the top and bottom of the liquefied layer. The development of pile damage at the interfaces between strong and weak soil layers is consistent with observations from case histories. (Hamada, 1992).

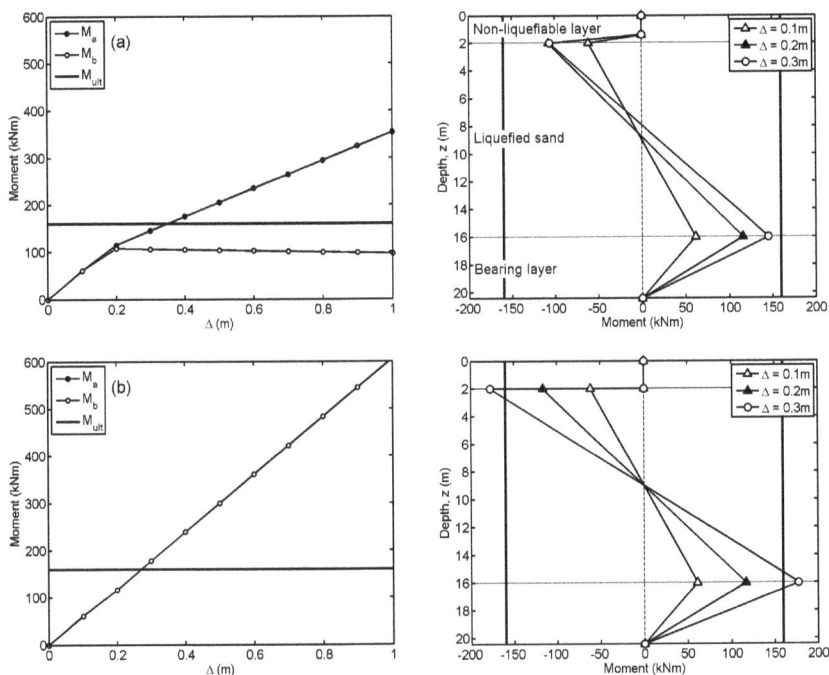

Fig. 5.10 Bending moments generated in (a) single pile and (b) pile within group (rigid cap).

5.4.4 Insight into effects of axial load on group response

As an example of the use of the LE model presented in section 5.4.2, a 2 × 2 group of piles with material properties representative of reinforced concrete is analysed in this section. The piles are of diameter 0.5m, length 14m, have a bending stiffness EI = 20MNm^2 and an ultimate moment = 160kNm. For the purposes of this example, p_l was taken to be zero. The results of this analysis are shown in Fig. 5.11. To produce these results, F_{crust} was computed for a range of values of Δ at a given axial load. The calculations were then repeated for different axial loads per pile between zero and the instability load P_{cr}.

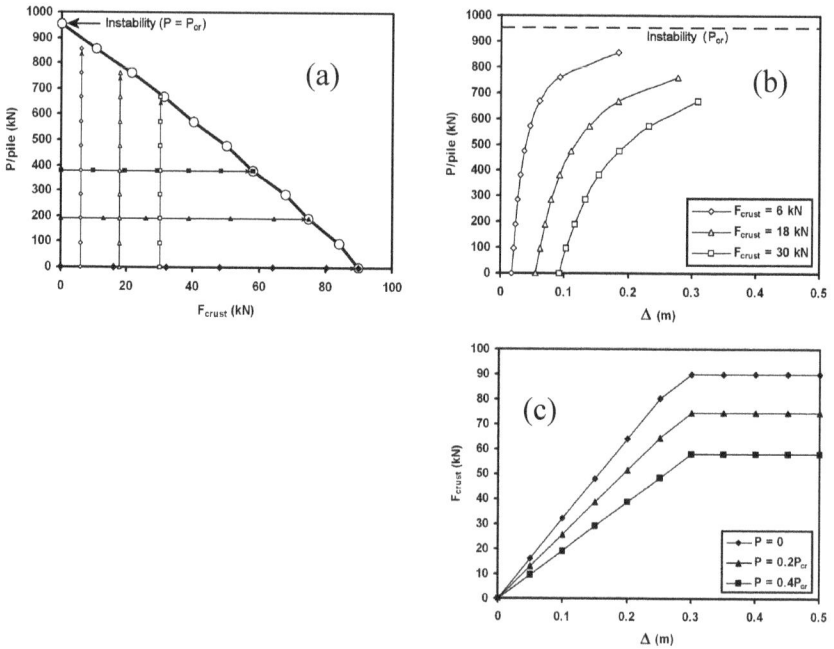

Fig. 5.11 Example showing the effect of axial load on structural response of pile group.

Figure 5.11a summarises the ultimate values of F_{crust} at bending failure of the pile. This gives an interaction diagram which demonstrates that, as mentioned previously, as the axial load in the piles is increased, the maximum crust load which can be sustained before pile failure reduces. For $P = P_{cr}$, the piles become unstable irrespective of any crustal loads. The effects of axial load on lateral displacements are shown in Fig. 5.11b for constant F_{crust}. This clearly shows that as well as reducing the maximum value of F_{crust} that can be tolerated (Fig. 5.11b), increased axial load will lead to larger displacements for given soil forcing. This effect becomes greater as the axial load approaches the instability load (P_{cr}). These two effects are summarised in Fig. 5.11c in which the pile group response is summarised in terms of the lateral force displacement response.

5.5 Validation of Effects of Axial Pile Load

In order to physically observe the effects indicated in Figure 5.11 and to validate the model proposed by Eqs. 5.5–5.9, dynamic centrifuge testing was conducted by Knappett (2006). In these tests a 2×2 pile group was considered in a three-layer soil profile inclined at an angle of $6°$ to the horizontal. Two separate and similar models were tested in which only the axial load applied to the piles was varied. The piles had bending properties similar to 0.5m diameter solid circular reinforced-concrete piles at prototype scale and with a centre-to-centre spacing of $5.6D_0$. The piles were 12.4m long, embedded in the dense sand bearing layer by $\sim 7D_0$. Salient properties and dimensions of the models are shown in Fig. 5.12.

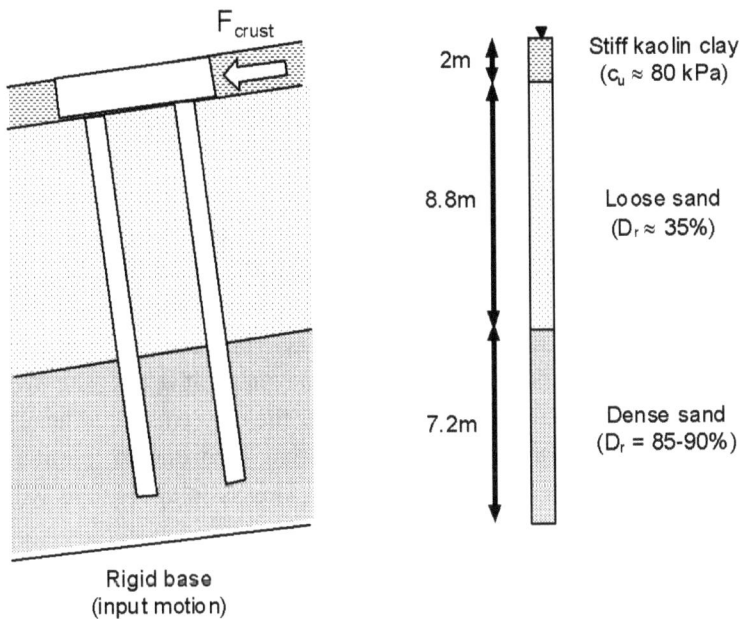

Fig. 5.12 Details of centrifuge tests conducted by Knappett (2006).

The total vertical load applied to the top of the pile cap in each model is presented in Table 5.3. Measurements of bearing pressure were made during the test and used to estimate the total amount of load carried by the pile cap. Nominal axial loads per pile were then estimated considering the remaining load to be distributed equally between the four piles.

Table 5.3 Axial loads acting in centrifuge tests.

Total axial load on pile cap (MN)	Estimated cap load (kN)	Nominal load per pile, P (kN)
1.80	280	380
4.80	2520	570

Each model was subjected to harmonic shaking with a peak input acceleration of ~0.3g. Pressures were measured on the upslope and downslope faces of the pile cap, which were used to estimate the total downslope forces applied by the non-liquefiable crust (F_{crust}). Lateral deflections of the pile cap were also measured. The measured behaviour is shown for the two different pile groups in Fig. 5.13. The original data has additionally been passed through a low-pass filter to remove the cyclic component. The measured monotonic (filtered) responses are compared to the computed response obtained using the LE method presented in Sec. 5.4.2. Due to the relative impermeability of the clay crust, the ultimate value of p_l was taken as -10kPa, which was assumed to be reached at a relative soil-pile displacement of 0.2m. The use of a negative (upslope) value of p_l accounts for the clay crust (and hence also the piles) moving further than the liquefied sand due to a water film forming between the crust and the sand. This is in agreement with the residual soil displacements observed after the tests.

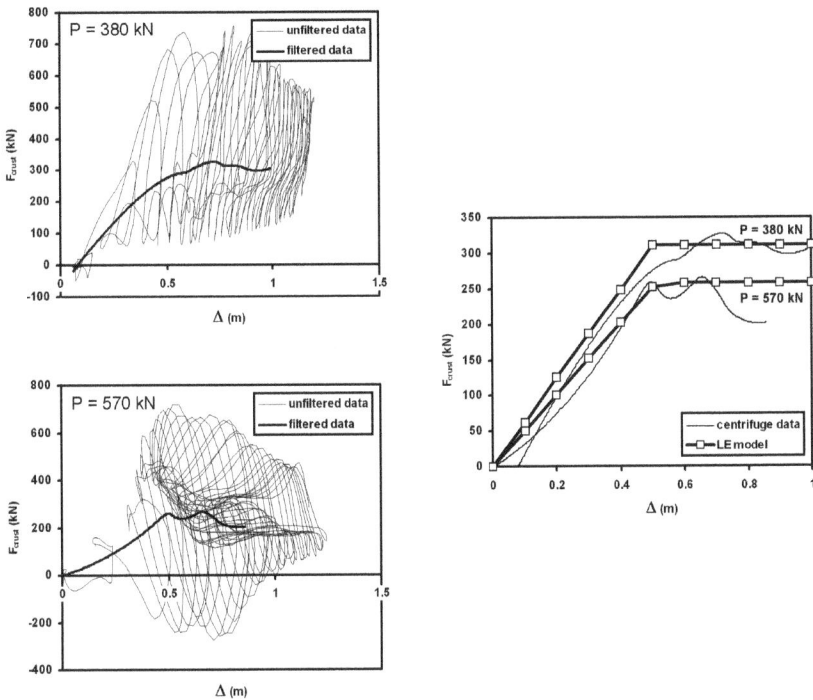

Fig. 5.13 Validation of LE model presented in Sec. 5.4.2 with centrifuge test data.

Both pile groups can be observed to reach a limiting value of F_{crust}, indicating that the piles yielded instead of the soil. This is consistent with the high yield strength of the clay used for the crustal layer. The agreement between the experimental data with the LE model, both in terms of the ultimate value of F_{crust} and the displacement at which this force is mobilised, is very close. This suggests that the LE model is able to adequately capture the essential features of the residual pile response in laterally spreading soils.

5.6 Recommendations for Designing Piles in Laterally Spreading Ground

This chapter has thus far considered the effects of axial load on both the lateral response (Δ) and ultimate lateral capacity of pile groups subjected to lateral kinematic forces in laterally spreading soil. For piles founded in cohesionless bearing layers, liquefaction-induced settlement and bearing failure may also occur, as discussed in Chapter 3 (Fig. 5.14). These considerations impose additional constraints on the maximum value of axial load that may safely be carried. In terms of the ultimate limiting state (ULS), this may be included as an additional line on the interaction diagram concept developed in Sec. 5.3.3, which changes the shape of the limiting load envelope. This is shown schematically in Fig. 5.15.

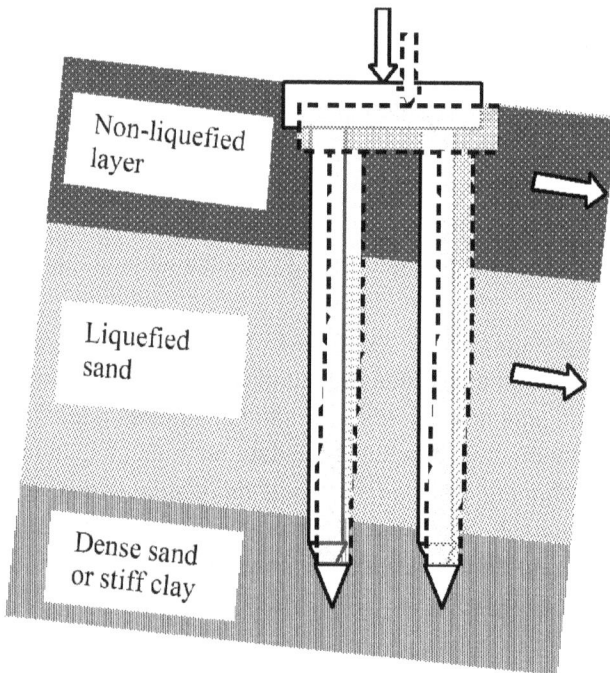

Fig. 5.14 Bearing failure of a pile group in laterally spreading soil (c.f. Sec. 1.6.2).

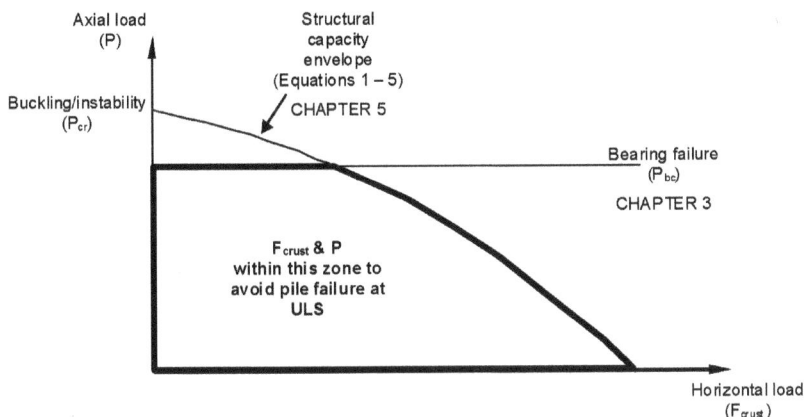

Fig. 5.15 Interaction diagram for pile group failure in laterally spreading soil.

For a given pile group design, an interaction diagram may be constructed using the methods presented in this chapter and Chapter 3. The maximum (ultimate) soil force at soil yield for the non-liquefiable layer ($F_{s,ult}$) may be found following the procedures outlined in Chapter 4. For regions of the interaction diagram for which F_{crust} at pile yield is greater than $F_{s,ult}$, the pile section and axial loads will be suitable as the crust will be unable to provide enough lateral force to cause pile failure. Conversely, for areas in which $F_{crust} < F_{s,ult}$, the piles require strengthening (increase in moment/shear capacity) so that $F_{crust} > F_{s,ult}$. Due to the shape of the interaction diagram, the first of these two conditions will tend to occur at lower axial loads, while the latter will occur at higher axial loads. This is shown schematically in Fig. 5.16. For a given pile section, there may therefore be a limiting axial load for the piles in order to ensure that they do not fail laterally. The value of this limiting load will depend on the pile properties (interaction diagram) and the crustal soil properties ($F_{s,ult}$). For strong crustal layers with low strength piles, the limiting axial load is likely to be very low. In these cases, larger piles will be required to provide a pile group which is more efficient at carrying axial load. An example of these effects is shown in Fig. 5.17.

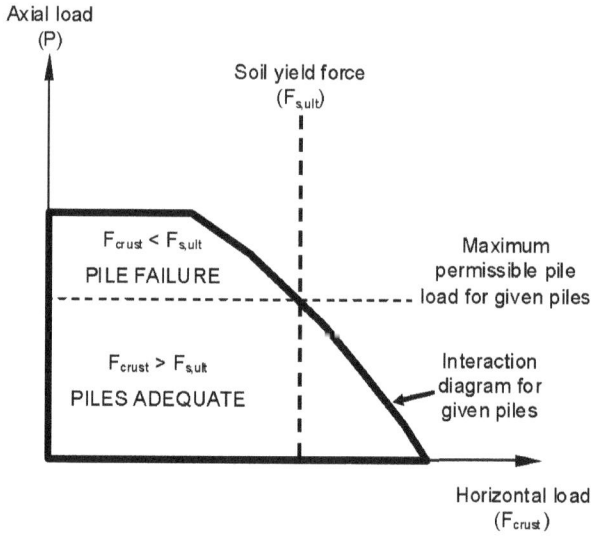

Fig. 5.16 Use of interaction diagrams for determining maximum permissible axial load to avoid lateral pile failure.

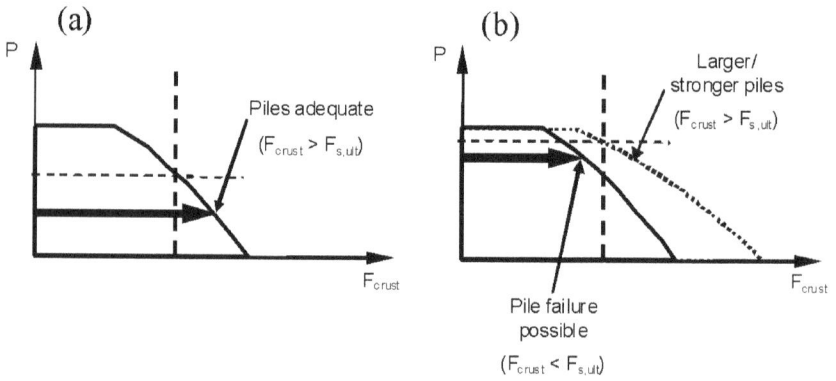

Fig. 5.17 Example of use of interaction diagrams for pile sizing: (a) at low axial load piles are adequate as crust will yield before piles and (b) at high axial load, larger/stronger piles are required to ensure that the crust yields before the piles.

The interaction diagram may also be used to determine a suitable pile section to satisfy lateral criteria, which may then be used to determine pile length and axial load against settlement/bearing capacity criteria using the methods outlined in Chapter 3. This is accomplished by selecting a pile size for which the limiting axial load at which $F_{crust} < F_{sult}$ is greater than P_{bc} or the limiting P for a given amount of settlement (c.f. Chapter 3).

The actual force mobilised during spreading may be less than $F_{s,ult}$, if the crust does not spread sufficiently to mobilise $F_{s,ult}$. The shape of the interaction diagram, however, ensures that a lower mobilised value of F_{crust} will only increase the permissible axial load which can be carried. Determination of the actual mobilised crustal load and pile group displacement for a given amount of soil displacement may be accomplished by constructing the $F_{crust}-\Delta$ and $y_{crust}-\Delta$ charts (c.f. Fig. 5.9). This is shown schematically in Fig. 5.18 for two different axial loads.

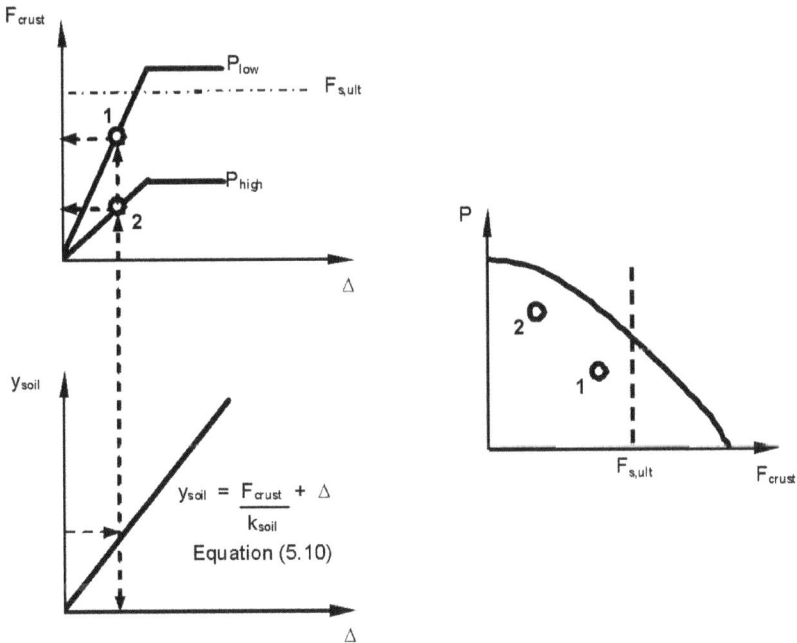

Fig. 5.18 Determination of mobilised crust load and group deflection for given y_{soil}.

y_{soil} may be predicted using empirical correlations or using a numerical free-field soil response analysis. Once y_{crust} is known, the $y_{crust}-\Delta$ chart is used to determine the compatible foundation displacement. With Δ known, the $F_{crust}-\Delta$ chart may then be used to determine the mobilised F_{crust} for a given P.

Chapter 6

Design Examples

6.1 Introduction

In this chapter, examples are presented to illustrate how the concepts discussed in Chapters 1 to 5 may be considered in an inclusive design method.

Examples will be based on idealised scenarios and use synthesised CPT data to fully illustrate how such data may be used to develop safe designs for piles when designing in liquefiable soils. Before embarking on the design of piles in liquefiable soils, some examples are presented on the static design of piles in the following section.

6.2 Design of Piles Under Static Loading

In this section, examples of the static design of piles are illustrated, based on the methods outlined in Chapter 1.

Design brief: Pile foundations are to be designed for a bridge pier. The axial load on the pile cap from the pier is expected to be 9.4MN. Initial borehole data suggests an 8m thick silty sand layer overlying a 22m deep, dense sand deposit. The water table is expected to be at the ground level. CPT tests were carried out using a 25.4mm diameter cone and the cone penetration resistance profile is presented in Fig. 6.1. The saturated unit weight of the silty sand layer is 17 kN/m^3 and that of the dense sand is 19kN/m^3. The critical state friction angle for both these soils is initially taken to be 32°.

The design earthquake is an L1 event (return period of 72 years), having a magnitude $M = 6$ generating a peak ground acceleration of 0.2g.

The foundation must remain safe (i.e. no foundation yielding or bearing capacity failure) and additionally, due to serviceability requirements in the superstructure, foundation movements must not exceed the following values:

Horizontal movement: 100mm
Vertical displacement: 75mm

To provide a starting point, 0.75m diameter (D_0) closed ended steel tubular piles will be considered with the properties shown in Table 6.1.

Table 6.1 Pile properties

Diameter (m)	Wall thickness (mm)	EI (MNm^2)	Yield stress (MPa)	Moment capacity, yield (kNm)
0.75	12	398	275	1800

6.2.1 Example 1: Preliminary design of piles under static loading

Initially, the pile is assumed to be of length 20m (L_p). Preliminary design can be carried out by using the methodology outlined in Chapter 1 using Eqs. 1.1 to 1.4 and Berezantzev *et al.*'s N_q factors.

6.2.1.1 End bearing

Pile base area is first computed as follows:

$$A_b = (\pi/4) \times 0.75^2 = \underline{0.442} \text{ m}^2. \tag{6.1}$$

Effective vertical stress at 20m depth (σ'_b) is calculated as follows, taking the unit weight of water to be 10kN/m^3. It is assumed that the pile is closed ended.

$$\sigma'_b = 17 \times 8 + 19 \times 12 - 20 \times 10 = \underline{164} \text{ kPa} \tag{6.2}$$

For a critical state friction angle of 32° for the dense sand, the value N_q can be taken as 40 from Fig.1.2. The base capacity of the pile can therefore be calculated using Eq. 1.2 as:

$$Q_b = 0.442 \times 164 \times (40-1) = \underline{2827} \text{ kN}. \tag{6.3}$$

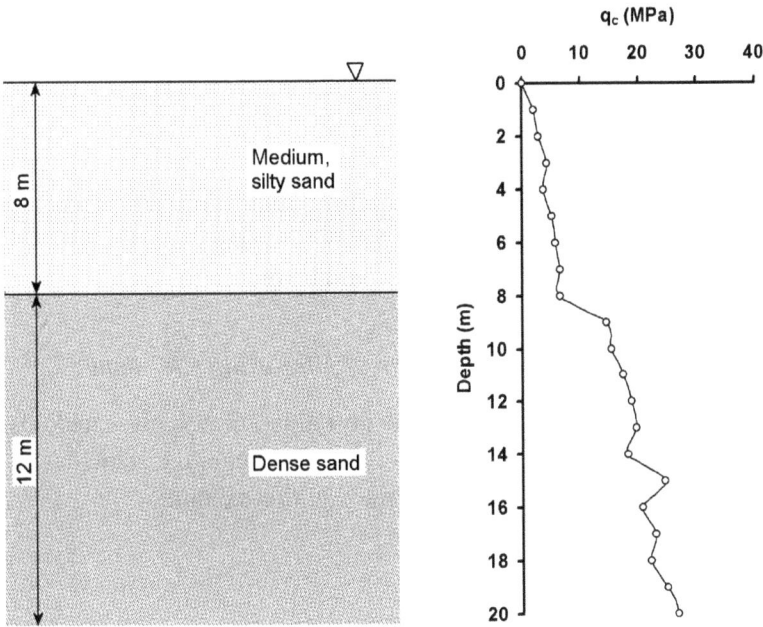

Fig. 6.1 Information from site investigation.

6.2.1.2 Shaft resistance

The friction angle between the pile and the sand is assumed to be 20°. Shear stress at any depth z in the medium silty sand and the dense sand layers is calculated using Eq. 1.3 and Table 1.1 as follows:

$$\tau_s = 0.5 \times (17-10)z \tan(20^o) = \underline{1.27z} \text{ kPa;} \tag{6.4}$$

$$\tau_s = 1.0 \times (19-10)z \tan(20^o) = \underline{3.28z} \text{ kPa.} \tag{6.5}$$

The shaft resistance can then be calculated using Eq. 1.4 as:

$$Q_s = \pi \times 0.75 \times \left(\int_0^8 1.27z \, dz + \int_8^{20} 3.28z \, dz \right) = \underline{1394} \text{ kN.} \tag{6.6}$$

Therefore, the pile capacity is obtained as:

$$Q_u = 2827 + 1394 = \underline{4221} \text{ kN.} \tag{6.7}$$

From this, the number of piles required to support the given axial load of 9.4MN can be calculated as:

$$N = \frac{9.4 \times 10^3}{4221} = \underline{2.2} .$$ (6.8)

As an initial design, choose five piles 750mm in diameter and 20m long to give an FOS of at least two against bearing capacity failure (for an FOS of at least three, seven piles would be required).

6.2.2 Example 2: Preliminary design of piles using CPT data

In Fig. 6.1 the CPT data from the site investigation was presented. Using the methods outlined in Chapter 1 and the CPT data, the static design of piles can be carried out. This is illustrated in this example.

6.2.2.1 End bearing

Using Eq. 1.7 or Fig. 1.3 (for a 25.4mm diameter cone), the end bearing capacity of the pile can be calculated from q_c at 20m depth.

$$\frac{q_b}{q_c} = 1 - 0.5 \log\left(\frac{750}{25.4}\right) = 0.265 \geq 0.13$$ (6.9)

$$q_b = 0.265 \times q_c = 0.265 \times 26.7 = \underline{7.08} \, \text{MPa}$$ (6.10)

Therefore, end bearing of the pile is calculated as:

$$Q_b = 7.08 \times 0.442 \times 1000 = \underline{3129} \, \text{kN}.$$ (6.11)

6.2.2.2 Shaft resistance

The shear stress variation with depth can be calculated using Eq. 1.8 following the MTD method suggested by Jardine and Chow (1996). As an initial approximation, the stress change at the pile tip due to the dilation of sand as it flows round the pile tip during pile driving may be taken to be negligible. As in example 1 (Sec. 6.2.1), the friction angle between the pile and sand is taken as 20°.

$$\tau_s = \left[\frac{q_c}{45} \left(\frac{\sigma'_{v0}}{100} \right)^{0.13} \left(\frac{0.75}{z} \right)^{0.38} \right] \tan 20^o \qquad (6.12)$$

Using Eq. 6.12 the shear stress between the pile and the soil can be calculated at any given depth as shown in Table 6.2.

Table 6.2 Calculation of shear stress variation with depth.

Depth (m)	Effective vertical stress σ'_{v0} (kPa)	q_c (MPa)	Shear stress* τ_s (kPa)
0	0	0.00	0.00
1	7	0.74	3.78
2	14	1.19	5.12
3	21	1.81	7.05
4	28	1.82	6.59
5	35	2.41	8.28
6	42	2.79	9.15
7	49	3.25	10.24
8	56	3.43	10.46
9	81	12.85	39.34
10	90	13.77	41.05
11	99	16.03	46.67
12	108	17.94	51.10
13	117	19.07	53.25
14	126	17.88	49.00
15	135	24.94	67.20
16	144	20.93	55.49
17	153	23.71	61.90
18	162	22.97	59.14
19	171	26.59	67.52
20	180	28.89	72.42

*calculated using Eq. 6.12.

The shaft resistance is then calculated by integrating the shear stress with depth and multiplying with the perimeter of the pile:

$$Q_s = \pi \times 0.75 \int \tau_s dz = \pi \times 0.75 \times 688.54 = \underline{1622} \text{ kN.} \qquad (6.13)$$

Therefore, the pile capacity is obtained as:

$$Q_u = 3129 + 1622 = \underline{4751} \text{ kN.} \tag{6.14}$$

The minimum number of piles required to carry an axial load of 9.4MN is:

$$N = \frac{9.4 \times 10^3}{4751} = \underline{2.0}. \tag{6.15}$$

From Eq. 6.15, four piles, 750mm in diameter and 20m long can support the required axial load for a static FOS of approximately two. The use of CPT test data has therefore led to a more economical design compared to the simple method presented in Example 1 in this case.

As an alternative to the MTD method proposed by Jardine and Chow (1996), the method presented by Randolph *et al.* (1994) could be used to calculate the shear stress as explained in Chapter 1. Following this method and assuming;

$$K_{max} = 0.015 \frac{q_c}{\sigma'_{vo}}$$

$$K_{min} = 0.4$$

$$\mu = 0.05$$

The following expression for shear stress may be written as:

$$\tau_s = \left[0.4 + \left(0.015 \frac{q_c}{\sigma'_{vo}} - 0.4 \right) e^{-0.05 z/0.75} \right] \sigma'_{vo} \tan 20^o. \tag{6.16}$$

Using Eq. 6.16 the ultimate shear stress between the pile and the soil can be calculated at any given depth as shown in Table 6.3.

As before, the shaft resistance is then calculated by integrating the shear stress with depth and multiplying with the perimeter of the pile:

$$Q_s = \pi \times 0.75 \int \tau_s dz = \pi \times 0.75 \times 688.16 = \underline{1621} \text{ kN.} \tag{6.17}$$

Therefore, the pile capacity is obtained as:

$$Q_u = 3129 + 1621 = \underline{4750} \text{ kN.} \tag{6.18}$$

This value is almost identical to that obtained by using the MTD method previously.

Table 6.3 Calculation of shear stress using Randolph *et al.* (1994) method.

Depth (m)	Effective vertical stress σ'_{v0} (kPa)	q_c (MPa)	q_c/σ'_{v0}	K_{max}	Shear Stress τ_s (kPa)
0	0	0.00	0.00	0.00	0.00
1	7	0.74	105.31	1.58	3.83
2	14	1.19	84.76	1.27	5.92
3	21	1.81	86.07	1.29	8.63
4	28	1.82	64.82	0.97	8.54
5	35	2.41	68.94	1.03	10.88
6	42	2.79	66.45	1.00	12.23
7	49	3.25	66.26	0.99	13.78
8	56	3.43	61.21	0.92	14.35
9	81	12.85	158.66	2.38	43.83
10	90	13.77	153.00	2.29	44.97
11	99	16.03	161.93	2.43	49.53
12	108	17.94	166.10	2.49	52.66
13	117	19.07	163.02	2.45	53.64
14	126	17.88	141.87	2.13	49.51
15	135	24.94	184.76	2.77	62.52
16	144	20.93	145.35	2.18	53.08
17	153	23.71	154.96	2.32	56.78
18	162	22.97	141.82	2.13	54.26
19	171	26.59	155.48	2.33	58.78
20	180	28.89	160.49	2.41	60.87

6.3 Inertial and Kinematic Loading on Piles in Level Ground

In this section simplified methods to estimate the inertial and kinematic loading on the pile group located in level ground will be presented. The soil profile that will be considered will be the same as in Fig. 6.1. The axial load on the pile group is 9.4MN as considered in the static design of the piles in Examples 1 and 2 above. The pile group has four piles in 2 × 2 formation with a pile spacing of $5D_0$. Based on the static design presented in earlier examples, the pile will be a steel tubular pile with an outer diameter of 750mm and a wall thickness of 12mm. The yield stress of steel is taken as 275MPa and the Young's modulus is taken as 210GPa. The flexural rigidity of each pile EI is 398MN m^2. The pile spacing will be 5×750mm = 2250mm.

Recall from Sec. 6.1 that the design earthquake is magnitude $M = 6$ generating a peak ground acceleration of 0.2g.

6.3.1 Example 3: Soil stiffness and natural frequency

Initially, it is assumed that the void ratio of the medium, silty sand layer and dense sand layer are 0.9 and 0.6 respectively.

The vertical effective stress at 4m depth can be seen from Table 6.3 to be 28kPa. Assuming the coefficient of earth pressure at rest K_o to be 0.46, the mean effective confining pressure p' is calculated as:

$$p' = \left(\frac{1 + 2K_o}{3} \right) \sigma'_{vo} = \underline{17.9} \text{ kPa.} \tag{6.19}$$

Using Eq. 2.2 proposed by Hardin and Drnevich (1972) as presented in Chapter 2, the small-strain shear modulus of the silty sand layer can be calculated as:

$$G_o = 100 \frac{(3 - 0.9)^2}{(1 + 0.9)} \sqrt{\frac{17.9}{1000}} = \underline{31.1} \text{ MPa.} \tag{6.20}$$

This small-strain shear modulus needs to be corrected for the cyclic shear strain amplitudes generated during the earthquake loading. It is well known that the shear modulus of soil degrades with increasing shear strain. Many researchers investigated this aspect. (Hardin and Drnevich, 1972; Dobry and Vucetic, 1991). High quality cyclic simple shear tests were carried out on non-plastic silts at different confining pressures and void ratios by Matesic and Vucetic (1998) and Doroudian and Vucetic (1999). Figure 6.2 is re-plotted from their experimental data for a silty sand sample with a PI = 0 and at a vertical stress of 70kPa. It shows the variation of normalised shear modulus and damping ratio with shear strain.

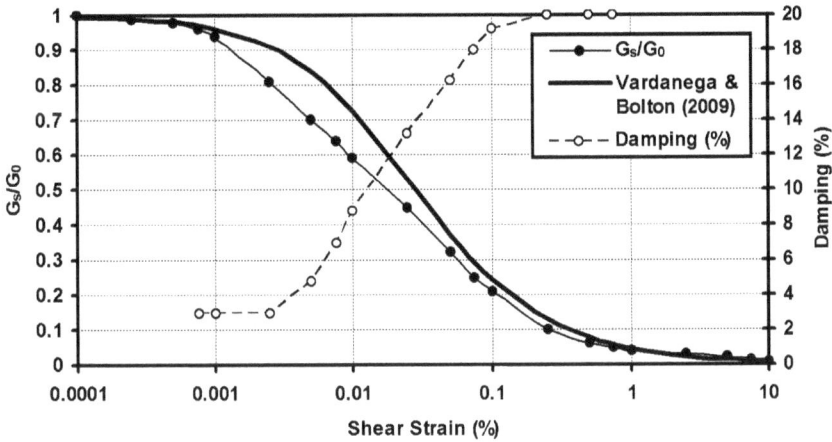

Fig. 6.2 Variation of normalised shear modulus and damping with shear strain.

The peak cyclic shear stress generated by an earthquake can be estimated as:

$$\tau_{max} = 0.65 \times \frac{a_g}{g} \times \sigma_{vo} \times r_d .$$ (6.21)

where r_d is a correction factor to allow for the deformable nature of the soil column being considered. This correction factor may be calculated as $r_d = (1 - 0.01z)$. (Iwasaki *et al.*, 1978).

Using Eq. 6.21 the maximum cyclic shear stress at mid-depth in the silty sand layer can be calculated as:

$$\tau_{max} = 0.65 \times \frac{0.2g}{g} \times 68 \times (1 - 0.01 \times 4) = \underline{8.49} \text{ kPa.}$$ (6.22)

In reality this value may be smaller as the earthquake acceleration at the base of the silt layer will normally be less than the peak ground acceleration. We also know that the shear stress is related to the shear strain as:

$$\tau_{max} = G_s \, \gamma .$$ (6.23)

Referring back to Fig. 6.2, we could assume a hyperbolic variation of the normalised shear modulus as shown in Eq. 6.24, following Hardin and Drnevich (1972) and with the modification suggested by Vardanega and Bolton (2009):

$$\frac{G_s}{G_o} = \frac{1}{\left(1 + \dfrac{\gamma}{\gamma_r}\right)^c} .$$ (6.24)

Using a reference strain γ_r of 0.02% and $c = 0.79$ fits the experimental data reasonably well as shown in Fig. 6.2. Using Eqs. 6.23 and 6.24, an expression can be obtained that involves only the shear strain on the RHS in Eq. 6.25:

$$\frac{\tau_{max}}{G_o} = \frac{\gamma}{\left(1 + \dfrac{\gamma}{\gamma_r}\right)^c} .$$ (6.25)

where γ is the absolute shear strain. Substituting the G_o and τ_{max} values from before gives:

$$\frac{\gamma}{\left(1 + \dfrac{\gamma}{2 \times 10^{-4}}\right)^{0.79}} = 2.73 \times 10^{-4} .$$ (6.26)

Equation 6.26 can be solved numerically and the value for peak cyclic shear strain is obtained as:

$$\gamma = \underline{0.144}\,\%.$$ (6.27)

From Fig. 6.2, for the above value of shear strain the shear modulus must be degraded to 20% of its initial value. Similarly at this strain level the damping in the silty sand will be about 19%. Therefore the secant modulus of the soil will be:

$$G_s = 0.2 \times 31.1 = \underline{6.22}\ \text{MPa}.$$ (6.28)

Using the shear modulus, the Young's modulus can be calculated from:

$$E_s = 2G_s\left(1+v\right). \tag{6.29}$$

Under undrained conditions, the Poisson's ratio v will be 0.5. Using this and Eq. 6.29, the Young's modulus of soil can be calculated as:

$$E_s = 3\times 6.22 = \underline{18.66}\,\text{MPa}. \tag{6.30}$$

The natural frequency of the silty sand layer can be calculated assuming the fundamental mode of vibration as shown by Eq. 2.1 shown in Chapter 2. Using the shear modulus calculated in Eq. 6.28, the shear wave velocity and the natural frequency can be calculated as:

$$v_s = \sqrt{\frac{6.22\times10^6}{1700}} = \underline{60.5}\,\text{m/s}; \tag{6.31}$$

$$f_n = \left(\frac{60.5}{4\times8}\right) = \underline{1.9}\,\text{Hz}. \tag{6.32}$$

It must be noted that if the silty sand layer suffers liquefaction during an earthquake, then this natural frequency can reduce significantly.

6.3.2 Example 4: Effective length and flexibility of the pile

6.3.2.1 Effective length of the pile

The concept of effective length was introduced in Chapter 2 in Sec. 2.3.2, which indicates the length of the pile that will participate in the inertial response during an earthquake. To determine the effective length of the pile, the Young's modulus of soil must be estimated at a depth D_0 which is the diameter of the pile. A parabolic variation of stiffness with depth within the silty sand layer will be assumed (Eq. 2.9).

The pile is a steel tubular pile with an outer diameter of 750mm and a wall thickness of 12mm. The Young's modulus of steel is 210GPa. The Young's modulus of the pile E_p needs to be corrected for the cross-section of the pile as explained in Sec. 2.9.2. Using Eq. 2.47:

$$E_{p\_corrected} = \frac{210}{0.75^4 \big/ \left(0.75^4 - 0.726^4\right)} = \underline{25.6}\,\text{GPa}. \tag{6.33}$$

In order to calculate the Young's modulus of soil at a depth of 0.75m, the effective mean confining stress at this depth must be calculated as shown before in Eqs. 6.19 and 6.20.

$$p' = \left(\frac{1 + 2 \times 0.46}{3}\right)5.25 = \underline{3.36}\,\text{kPa};\qquad(6.34)$$

$$G_o = 100\frac{(3 - 0.9)^2}{(1 + 0.9)}\sqrt{\frac{3.36}{1000}} = \underline{13.45}\,\text{MPa}.\qquad(6.35)$$

Using the same stiffness degradation assumed in Example 4, the degraded shear modulus can be calculated as:

$$G_s = 0.2 \times 13.45 = \underline{2.69}\,\text{MPa}.\qquad(6.36)$$

The Young's modulus at a depth of 0.75m can be calculated assuming undrained conditions as:

$$E_{sD} = 3 \times 2.69 = \underline{8.07}\,\text{MPa}.\qquad(6.37)$$

The effective length L_{ad} can then be calculated using Eq. 2.9 as:

$$L_{ad} = 2 \times 0.75\left(\frac{25.6 \times 10^9}{8.07 \times 10^6}\right)^{0.22} = \underline{8.84}\,\text{m}.\qquad(6.38)$$

This suggests that 8.84m length of the pile will participate in the inertial response. This extends below the silty sand into the dense sand layer by 0.84m. It must noted that the effective length above will be lower than calculated in Eq. 6.38 as the dense sand will be stiffer, so the length of the pile in the dense sand layer that participates in the inertial response will be less than 0.84m.

6.3.2.2 Flexibility of the pile

The flexibility of the pile relative to the silty sand layer can be estimated by using Eq. 2.11. For using this equation, a value for the gradient of soil modulus k needs to be chosen. The range of values k can take is between 200 and 2000kN/m^3. Using the lower value for k:

$$T_u = \left(\frac{E_p I_p}{k}\right)^{0.2} = \left(\frac{398 \times 10^6}{200 \times 10^3}\right)^{0.2} = \underline{4.569}\,\text{m};\tag{6.39}$$

$$Z_L = \left(\frac{L}{T_u}\right) = \left(\frac{20}{4.569}\right) = \underline{4.38}\,\text{m}.\tag{6.40}$$

As $2.5 < Z_L < 5$, the pile should be semi-flexible. Using the higher value for k:

$$T_l = \left(\frac{398 \times 10^6}{2000 \times 10^3}\right)^{0.2} = \underline{2.88}\,\text{m};\tag{6.41}$$

$$Z_u = \left(\frac{L}{T_l}\right) = \left(\frac{20}{2.88}\right) = \underline{6.94}\,\text{m}.\tag{6.42}$$

As $Z_u > 5$, the pile should be flexible. So the true behaviour of the pile will be between semi-flexible and flexible depending on the actual value of the gradient of soil modulus k.

6.3.3 Example 5: Inertial loading on the pile

The peak ground acceleration is 0.2g as before. It is assumed that this will be the acceleration at the base of the pile group. The amplification of this acceleration within the pile group needs to be calculated. Also, it is assumed that the axial load of 9.4MN is due to a concentrated (lumped) mass of the superstructure of 940 tons at a height of 10m above the pile cap.

The stiffness of each pile in the pile group is determined by considering Eurocode 8 provisions as discussed in Sec. 2.9.2 of Chapter 2. First the ratio of pile stiffness relative to the soil stiffness is calculated using the values for pile and soil stiffnesses from Example 4:

$$\left(\frac{E_P}{E_{sD}}\right) = \left(\frac{25.6 \times 10^9}{8.07 \times 10^6}\right) = \underline{3172.2}\,.\tag{6.43}$$

Using Table 2.3 and assuming a square root variation of stiffness with depth for the silty sand layer, the horizontal stiffness is calculated as:

$$K_h = 0.79 D E_{sD} \left(\frac{E_P}{E_{sD}} \right)^{0.28}$$ (6.44)

$$= 0.79 \times 0.75 \times 8.07 \times 10^6 \times (3172.2)^{0.28} = \underline{45.7}\ \text{MN/m}.$$

Similarly, the rotational stiffness is calculated as:

$$K_M = 0.15 D^3 E_{sD} \left(\frac{F_P}{E_{sD}} \right)^{0.77}$$ (6.45)

$$= 0.15 \times 0.75^3 \times 8.07 \times 10^6 \times (3172.2)^{0.77} = \underline{253.6}\ \text{MNm/rad}.$$

The coupling stiffness K_{hM} is calculated as:

$$K_{hM} = -0.24 D^2 E_{sD} \left(\frac{E_P}{E_{sD}} \right)^{0.53}$$ (6.46)

$$= -0.24 \times 0.75^2 \times 8.07 \times 10^6 \times (3172.2)^{0.53} = \underline{-78.1}\ \text{MN}.$$

The equivalent horizontal and rotational stiffnesses that involve the coupling stiffness K_{hM} can be written as shown in Eqs. 2.6 and 2.7 in Chapter 2 as:

$$K_{h\_eq} = \frac{K_h K_M - K_{hM}^2}{K_M - e K_{hM}}$$ (6.47)

$$K_{\theta\_eq} = \frac{K_h K_M - K_{hM}^2}{K_h - K_{hM}/e}$$ (6.48)

where e is the eccentricity.

In this problem we have a pile cap into which the four piles are fixed. Therefore, the rotation θ may be taken as zero within the elastic range i.e. no plastic hinges are formed at the pile heads. Using this condition, the rotational stiffness in Eq. 6.48 must be infinity ($K_{\theta\_eq} = \infty$). This is only possible if the denominator in Eq. 6.48 is zero, i.e.:

$$K_h - \frac{K_{hM}}{e} = 0 \quad \Rightarrow \quad e = \frac{K_{hM}}{K_h}$$ (6.49)

$$e = \frac{-78.1}{45.7} = \underline{-1.71} \,. \tag{6.50}$$

Substituting this and the values of stiffnesses from before into Eq. 6.47, we can write:

$$K_h = \frac{45.7 \times 10^6 \times 254 \times 10^6 - (-78.1 \times 10^6)^2}{253.6 \times 10^6 - (-1.71 \times -78.1 \times 10^6)} = \underline{45.7} \text{ MN/m} \tag{6.51}$$

The natural frequency of the pile group can then be calculated using the horizontal stiffness of each pile calculated in Eq. 6.51.

$$f_p = \frac{1}{2\pi} \left(\sqrt{\frac{4 \times 45.7 \times 10^6}{940 \times 10^3}} \right) = \underline{2.22} \text{ Hz.} \tag{6.52}$$

The natural frequency of the pile group is 2.22Hz and its natural period is therefore 0.45 seconds. If we approximate the pile group as a single degree of freedom system, we can use the design spectrum for sand from Eurocode 8 shown in Fig. 6.3. This design spectrum has been plotted for a soil factor of 1.35 and a damping of 20%. The damping for the silty sand layer at the strain level of 0.14% is nearly 20% as seen from the experimental data presented in Fig. 6.2.

From Fig. 6.3, it can be seen that for a time period of 0.45 seconds, the spectral acceleration is $2.13 \times a_g$. The response acceleration of the pile group will therefore be $2.13 \times 0.2g = 4.2 \text{m/s}^2$.

The horizontal inertial load on the pile group will be:

$$H = 940 \times 10^3 \times 4.2 = \underline{3.95} \text{ MN.} \tag{6.53}$$

The moment load on the pile group will be:

$$M = 940 \times 10^3 \times 10 \times 4.2 = \underline{39.5} \text{ MNm.} \tag{6.54}$$

The peak horizontal displacement δ_h during the inertial response of the pile group can now be calculated using the stiffnesses calculated in Eq. 6.51 as:

$$\delta_h = \frac{3.95}{45.7 \times 4} = \underline{0.022} \text{ m.} \tag{6.55}$$

The above response will change if the silty sand layer were to liquefy.

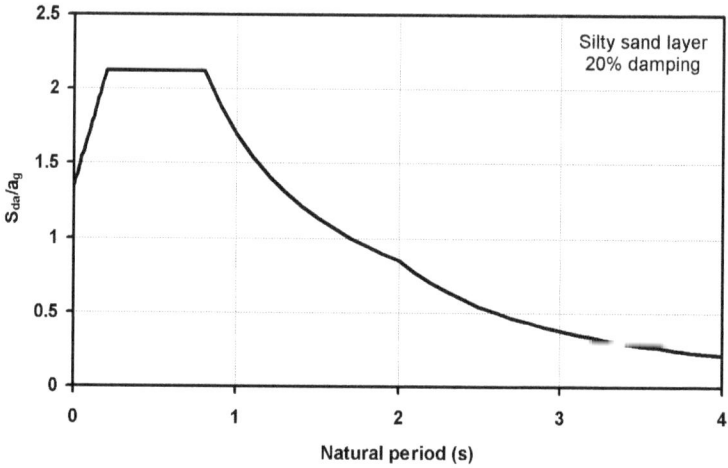

Fig. 6.3 Eurocode 8 design spectrum.

6.3.4 Example 6: Kinematic interaction

The kinematic interaction between the pile and the soil can be analysed by using the concept of the Interaction Factor I_u and a dimensionless factor F, introduced in Sec. 2.4 of Chapter 2.

The ratio of the stiffness of the pile to the stiffness of silty sand was calculated in the earlier examples. Similarly, values for the natural frequency of the silty sand layer (in the absence of liquefaction) and the pile group were obtained in earlier examples. Therefore, the dimensionless factor F for a parabolic variation of stiffness with depth can be calculated using Eq. 2.15 as:

$$F = \left(\frac{f_p}{f_n}\right)\left(\frac{E_p}{E_{sD}}\right)^{0.16}\left(\frac{L}{D}\right)^{-0.35}$$

$$= \left(\frac{2.2}{1.9}\right)(3172.2)^{0.16}\left(\frac{20}{0.75}\right)^{-0.35} = \underline{1.345}$$

(6.56)

The interaction factor can be calculated using values for the constants for parabolic variation of silty sand stiffness from Table 2.2 and substituting in Eq. 2.16 as:

$$I_u = 3.64 \times 10^{-6} \times 1.345^4 - 4.36 \times 10^{-4} \times 1.345^3$$
$$+ 6 \times 10^{-3} \times 1.345^2 + 1 \qquad (6.57)$$
$$= \underline{1.01}$$

If we assume that the earthquake results in a horizontal displacement u_o of 92mm at the surface of the silty sand layer, then the top of the pile group will have a displacement of;

$$u_p = I_u \times u_o = 1.01 \times 92 = \underline{93}\,\text{mm}. \qquad (6.58)$$

The silty sand layer is likely to have a response that will depend on its natural frequency of 1.9Hz. If the horizontal displacement at the surface of the silty sand layer varies with frequency as shown by u_o in Fig. 6.4, the pile group response will be as shown by u_p in Fig. 6.4.

If the natural frequency of the pile group is varied, the changes in the interaction factor between the pile group and the silty sand layer can be plotted as shown in Fig. 6.5.

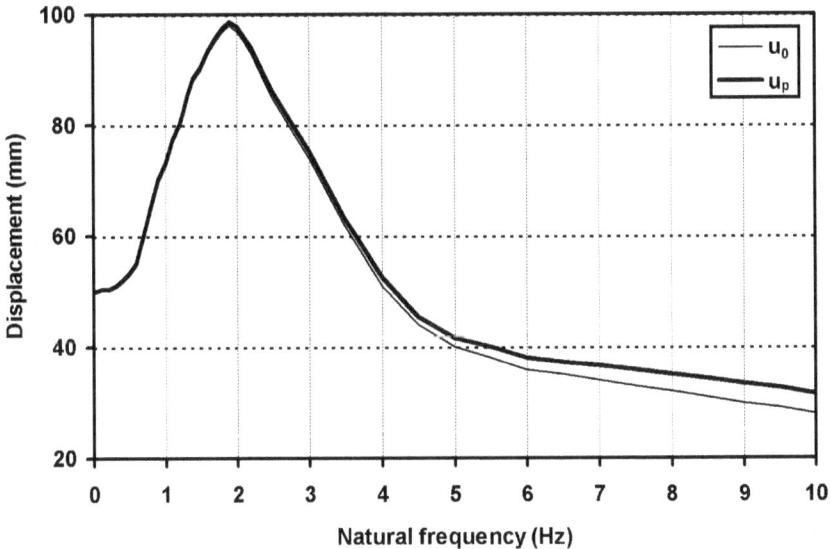

Fig. 6.4 Free field and pile head response due to kinematic interaction.

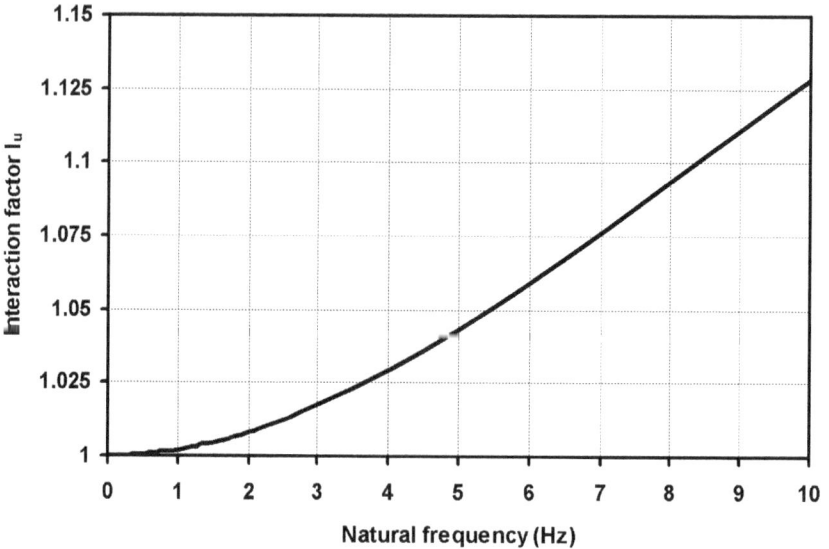

Fig. 6.5 Variation of Interaction Factor I_u with natural frequency of the pile group.

6.4 Design of Piles in Level Liquefiable Ground

The following example extends Examples 1–6 by considering the effects of liquefaction. The foundation for the bridge pier which was designed to static criteria in Sec. 6.2 will be considered here, accounting for the effects of soil liquefaction. The foundation is sited in level ground profile shown in Fig. 6.1 where there is a negligible risk of lateral spreading. Seismic CPT data has been collected close to the proposed location of the foundation. The seismic event considered is a magnitude M = 6 earthquake with 0.2g peak ground acceleration as in Sec. 6.3.

6.4.1 Example 7: Determination of liquefaction potential from CPT data

Figure 6.6 shows seismic CPT data for the site. Friction sleeve measurements (not shown) have verified that the soil profile consists of two main layers. The upper layer, a slightly silty sand, extends from the surface to 8m below ground level (BGL), and a clean sand layer extends below this. The saturated soil unit weights γ_s have been confirmed as 17kN/m^3 (average) within the upper layer, and 19kN/m^3 (average) within the lower layer. Relative density D_r was computed from the cone resistance according to Kulhawy and Mayne (1990):

$$D_r^2 = \frac{q_{c1}}{350}.$$ (6.59)

where q_{c1} is the normalised CPT resistance corrected for vertical effective stress given by:

$$q_{c1} = \frac{q_c}{p_a}\left(\frac{p_a}{\sigma'_{v0}}\right)^{0.5}.$$ (6.60)

where p_a is atmospheric pressure (= 100kPa). Peak secant friction angles ϕ_{pk} were computed after Robertson and Campanella (1983):

$$\tan\phi_{pk} = \frac{1}{2.68}\left[\log\left(\frac{q_c}{\sigma'_{v0}}\right) + 0.29\right].$$ (6.61)

Values of the critical state friction angle (ϕ_{cs}) for both soils were obtained by applying the relative dilatancy index of Bolton (1986) to the peak friction angles, with average values taken for each soil layer. The water table was observed to be just below the ground surface.

Before pile design can be carried out, the liquefaction hazard must be evaluated. Calculation of limiting axial loads in liquefiable soils, as presented in Chapter 3, requires knowledge of the excess pore pressure ratio with depth and not just the depth within which full liquefaction is triggered. This may be found using simple site response methods; however, often these methods require the input of soil parameters governing contractile and dilative characteristics of the soil, which may not be known without an extensive programme of triaxial testing. A simple method of estimating the maximum r_u profile induced by a given earthquake is presented here, based on the popular methodology outlined originally by Seed and Idriss (1971) and most recently modified by Idriss and Boulanger (2004).

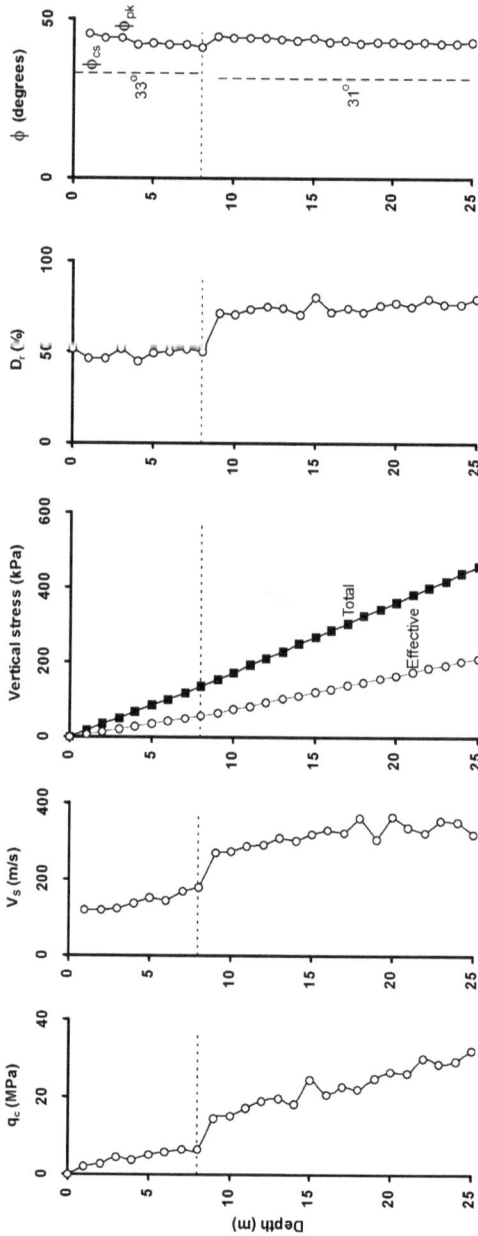

Fig 6.6 Seismic CPT data and derived soil parameters.

In this method, a Cyclic Resistance Ratio (CRR) is computed at each depth based on the soil properties. CRR represents the threshold cyclic shear stress at which full liquefaction is triggered. Based on Idriss and Boulanger (2004), CRR may be determined directly from the cone resistance for soils with <5% fines content using:

$$CRR = \exp\left[\frac{q_{cn}}{540} + \left(\frac{q_{cn}}{67}\right)^2 - \left(\frac{q_{cn}}{80}\right)^3 + \left(\frac{q_{cn}}{114}\right)^4 - 3\right]. \tag{6.62}$$

where $q_{cn} = q_c/p_a$. The remaining constants in the equation have been found based on an empirical fit to a large database of reported case histories where full liquefaction both did and did not occur.

The Cyclic Stress Ratio (CSR) is then computed at each depth which is a measure of the peak cyclic shear stresses induced by the design earthquake. This may be found using:

$$CSR = 0.65 \left(\frac{a_{max}\sigma_{v0}}{\sigma'_{v0}}\right) \frac{r_d}{MSF}. \tag{6.63}$$

where a_{max} is the peak ground acceleration (0.2g in this case) and r_d is given by:

$$r_d = \exp\left(\alpha + \beta M\right) \tag{6.64}$$

where

$$\alpha = -1.012 - 1.126\sin\left(\frac{z}{11.73} + 5.133\right), \tag{6.65}$$

z is the depth below the ground surface and

$$\beta = 0.106 + 0.118\sin\left(\frac{z}{11.28} + 5.142\right). \tag{6.66}$$

MSF is the earthquake Magnitude Scaling Factor and may be found from:

$$MSF = 6.9\exp\left(\frac{M}{4}\right) - 0.058. \tag{6.67}$$

Fig. 6.7 shows the results of these evaluations for the CPT data shown in Fig. 6.2. CRR plots as a curve which varies with the normalised cone resistance (q_{cn}), while calculated values of CSR are shown as individual data points. Data points falling above the line indicate the occurrence of full liquefaction.

It can be seen from Fig. 6.7 that in this example, the upper layer of sand is expected to fully liquefy as these points all plot above the CRR line. Below this depth, the data points plot beneath the line. The depth of full liquefaction ($r_u = 1$) is therefore 8m. This does not mean that no excess pore pressure will develop in the dense lower layer, however, only that full liquefaction conditions will not be reached. As in Chapter 3, it is possible to use a simple bilinear profile of excess pore pressure with depth to estimate the r_u profile. The maximum excess pore pressures are assumed to be constant with depth below the depth of full liquefaction (8m) and equal to the value at 8m depth (the bottom of the liquefiable layer), which leads to the variation of r_u as shown on the right of Fig. 6.7.

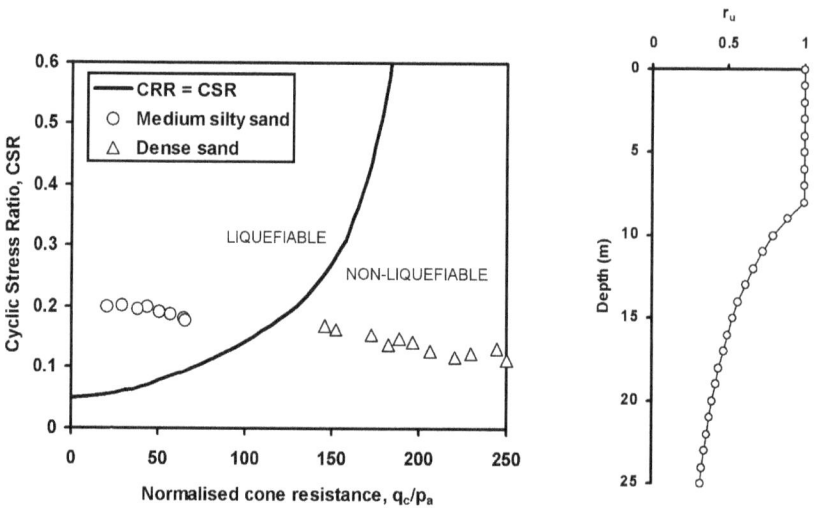

Fig. 6.7 Evaluation of liquefaction potential and extrapolation of r_u profile.

6.4.2 Example 8: Pile sizing based on liquefaction considerations

Now that the liquefaction potential has been evaluated in Example 7, the charts presented in Chapter 3 may be used to determine the range of suitable pile lengths for a given pile diameter. This procedure is based around consideration of axial failure modes, i.e. bearing and instability failures (ultimate limit state, ULS) and settlement considerations (serviceability limit state, SLS). For clarity, only a single pile size and material will be presented in this example; however, the process may be repeated for a range of different pile sizes which may then be evaluated against lateral load/displacement conditions so that an optimum design may be found.

Bearing in mind that a laterally stiff foundation is generally beneficial for minimising lateral displacements, a 0.75m diameter steel tubular pile section is selected, which will be closed-ended. The properties of the section considered are given in Table 6.1. To obtain a suitable range of pile lengths, a preliminary check against bearing failure and instability is conducted, as detailed in Sec. 3.7. The design chart has here been re-plotted as Fig. 6.8 in terms of maximum and minimum EI values so as to be more applicable in general design. From the liquefaction evaluation (Sec. 6.4.1), the depth of full liquefaction (z_L) was found to be 8m so that the demand curve is given by Eq. 6.28:

$$r_{u,base} = \frac{8}{L_p} . \tag{6.68}$$

The stiff steel tubular piles chosen are closer to the maximum EI such that the pile length may be chosen from the following range:

$$11\text{m} < L_p < 21 \text{ m}.$$

Note that with the availability of detailed CPT data, the minimum length (representing the onset of bearing failure) given by Fig. 6.8 is only a rough estimate. The minimum pile length may be found directly by considering the maximum loads that may be carried by the pile as determined from the CPT methods detailed in Chapter 1. Three limiting criteria will be considered:

1. Under static conditions a certain FOS must be achieved (here taken to be two). Detailed examples of design to this criterion has already been demonstrated in Examples 1 and 2.

2. FOS must be sufficiently high for liquefaction-induced bearing failure to be avoided. This is achieved if

$$FOS \geq \frac{1}{\alpha_{ult}\left(1 - r_{u,base}\right)^{\frac{3-\sin\phi}{3(1+\sin\phi)}} - \alpha_{ult} + 1} \quad (6.69)$$

3. FOS must be sufficiently high to avoid liquefaction-induced settlement $>0.1D_0$ (i.e. 75mm). This is achieved if

$$FOS \geq 1 + 5.5\left(r_{u,base}\right)^{3.5} \quad (6.70)$$

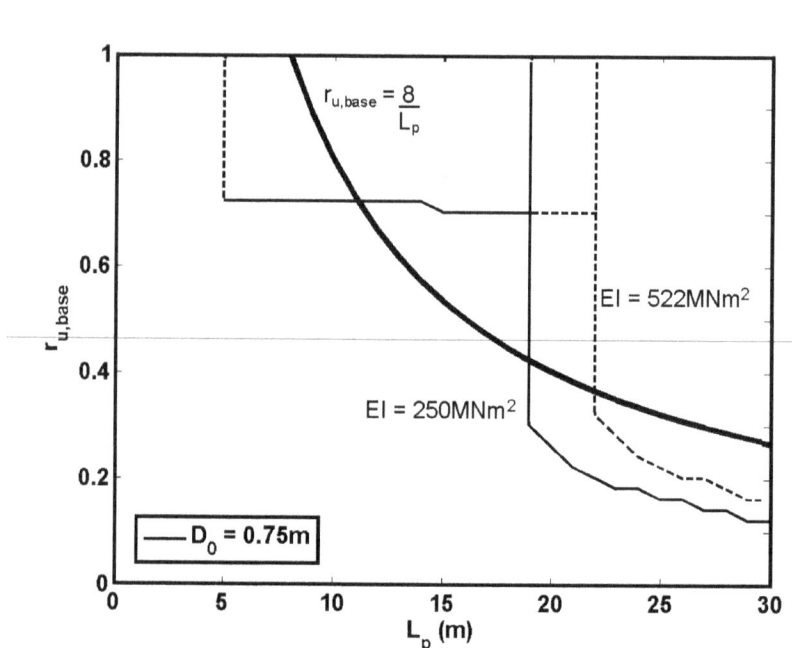

Fig. 6.8 Estimation of suitable pile length range to avoid bearing and instability failure (ULS).

For a pile founded at any depth within the soil layer, the FOS for each of the three criteria may be computed at each depth based on the profile of r_u determined in Fig. 6.7. The maximum FOS of the three values is then selected for determination of the minimum pile design load (P). This requires knowledge of the ultimate pile capacity (P_{ult}), which is found in this example using the MTD method (c.f. Example 2). The minimum number of piles (N) required in the foundation to produce an acceptable design is then given by:

$$N = \frac{P_{total} \cdot FOS_{max}}{P_{ult}}$$ (6.71)

For the example presented here, the FOS under each of the criteria are shown on the left of Fig. 6.9, and the associated minimum number of piles on the right of the same figure.

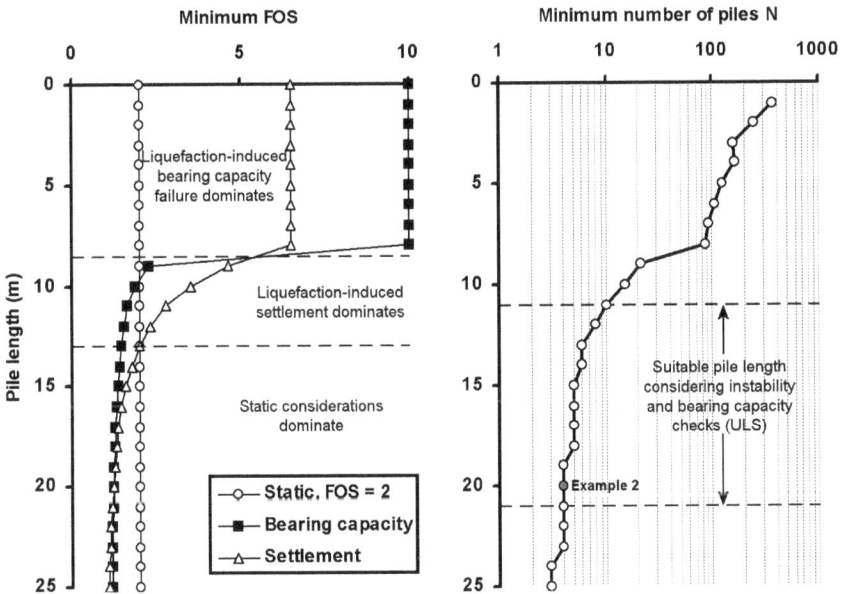

Fig. 6.9 Evaluation of minimum FOS under different criteria for piles of different lengths and associated minimum number of piles required in the group for FOS = 2 ($D_0 = 0.75$m).

From Fig. 6.9 the following conclusions may be drawn about a suitable founding depth for the piles:

- If piles are founded in the liquefiable layer, liquefaction-induced bearing failure is the controlling factor. The high FOS required to avoid this means that for these lengths the piles are very inefficient so that an unreasonable number would be required for a safe design.
- For 8m $<L_p$ <12m, the liquefaction-induced settlement criteria dominates. The piles are still being used inefficiently (under static conditions FOS could be as high as five). However, a reasonable minimum number of piles are required in the group. If, due to horizontal loading considerations, it is desirable to use a large number of piles, Fig. 6.9 shows that it may be reasonable to reduce the pile lengths to provide a more efficient design.
- For piles longer than 12m, the static FOS condition is controlling, so that piles longer than this should not suffer any adverse *axial* effects of liquefaction. In any case, the piles must not be longer than 21m to satisfy the instability constraints.

The result of the analysis thus far is a range of suitable pile properties which may be finalised by considering lateral loading effects and horizontal displacement criteria. In this case, the pile design is summarised as follows:

Table 6.4 Summary of pile group design following axial failure checks.

Diameter (m)	EI (MNm^2)	Moment capacity, yield (kNm)	Pile length (m)	Number of piles in group
0.75	398	1800	$12 < L_p < 21$	$N \geq 4$ (for all L_p)

The foundation design that was developed in Examples 1–6 satisfies all of the criteria in Table 6.4. It only remains, therefore, to check the horizontal response of the foundation when liquefaction is taken into account – this will be considered in Example 9.

6.4.3 Example 9: Inertial response in level liquefied ground

In this example, the inertial response in the liquefied ground is found and compared with the case of no liquefaction considered in Examples 4–6.

The liquefied soil is conservatively assumed to have approximately zero stiffness ($E_{sD} \approx 0$) for determination of the inertial load. With this approximation, the foundation will behave as a partially embedded pile with a free-standing length of 8m. The relative pile-soil stiffness in the dense sand is found using Eq. 2.11 with $k \approx 9000$ kN/m^3 for sand at $D_r = 75\%$:

$$T_u = \left(\frac{E_p I_p}{k}\right)^{0.2} = \left(\frac{398 \times 10^6}{9000 \times 10^3}\right)^{0.2} = \underline{2.13}\,\text{m.} \tag{6.72}$$

The fixity depth within the dense layer is then found after Davisson (1970). If the depth of the liquefied layer is less than T_u, the fixity depth is given as $2.2T_u$. If the liquefied layer is deeper than T_u, the fixity depth is $1.8T_u$ (as in this case 8m > 2.13m).

$$L_f = 1.8T_u = \underline{3.83}\,\text{m} \quad (\approx 5D_0) \tag{6.73}$$

The lateral stiffness of the 2×2 pile group in the liquefied ground can then be estimated considering the foundation to behave as a sway frame. Using basic structural beam theory, the lateral stiffness of a single pile within the group is therefore given by:

$$K_{h\_eq} = \frac{12EI}{\left(L_{p,layer1} + L_f\right)^3} = \frac{12 \times 398 \times 10^6}{\left(8 + 3.8\right)^3} = \underline{2.9}\,\text{MN/m.} \tag{6.74}$$

The natural frequency of the piled foundation is then found as in Example 5:

$$f_p = \frac{1}{2\pi}\left(\sqrt{\frac{4 \times 2.9 \times 10^6}{940 \times 10^3}}\right) = \underline{0.56}\,\text{Hz.} \tag{6.75}$$

The natural frequency of the pile group is 0.56Hz and its natural period is therefore 1.79 seconds. If we approximate the pile group as a single degree of freedom system as in Example 5, we can use the design spectrum for sand from Eurocode 8 shown in Fig. 6.3. The use of this spectrum assumes that the damping in the liquefied soil will be approximately 20%, which is reasonable as the soil will undergo large shear strain in the liquefied state.

From Fig. 6.3, it can be seen that for a time period of 1.79 seconds, the spectral acceleration is $0.9 \times a_g$. The response acceleration of the pile group in the liquefied ground will therefore be $0.9 \times 0.2g = 1.8\text{m/s}^2$. The horizontal inertial load on the pile group will be:

$$H = 940 \times 10^3 \times 1.8 = \underline{1.69}\,\text{MN}. \tag{6.76}$$

The peak horizontal displacement during the inertial response of the pile group can now be calculated using the equivalent horizontal stiffness calculated in Eq. 6.74 as:

$$\delta_h = \frac{1.69}{2.9 \times 4} = \underline{0.146}\,\text{m}. \tag{6.77}$$

By comparing the results of this example with Example 5, it can be seen that liquefaction of the upper soil layer

- reduces the lateral stiffness of the foundation substantially;
- lengthens the natural period of the foundation;
- reduces acceleration at the top of the foundation (which will become the input acceleration for the superstructure); and
- increases the lateral response of the foundation substantially.

6.5 Design of Piles in Sloping Liquefiable Ground

6.5.1 Example 10: Pile group in two-layer soil profile subject to lateral spreading

The soil profile considered in the previous examples (and shown in Fig. 6.1) is now assumed to gently slope at an angle of 3° to the horizontal. The pile group which has been designed over the previous examples will now be checked against lateral spreading of the soil. The pile cap is assumed to be 6m × 6m in plan and 1.5m thick. This would permit a pile-to-pile spacing within the group of ~$5D_0$ which is sufficiently large to neglect pile-soil-pile interaction effects.

Following the recommendations of Chapter 4, limiting lateral earth pressures will be used, rather than a displacement-based approach. This is particularly suitable in this case considering the difference in relative stiffness between the pile group and the liquefied soil, which would suggest that large relative soil-pile displacement will occur. A lateral pressure of 20kPa will be assumed to act laterally downslope on each pile and on the upslope face of the pile cap. Shear forces on the sides and underside of the pile cap are neglected. The foundation and the forces acting on it are shown in Fig. 6.10.

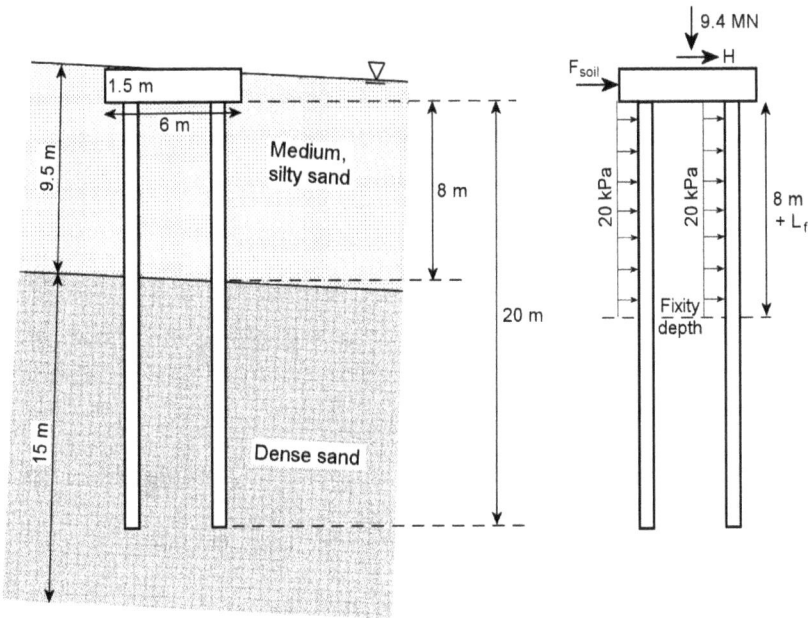

Fig. 6.10 Layout of pile group in two-layer soil profile, sloping at 3°.

By combining Eqs. 5.5–5.9 it is possible to obtain an expression for the expected pile group displacement in terms of the applied forces/pressures:

$$\delta_h = \frac{\left[2\left(F_{cap} + H\right) + Np_L D_0 L\right]L}{2NP\left(2f_\Delta - 1\right)}. \tag{6.78}$$

In Eq. 6.78, the parameter (F_{cap} + H) replaces F_{crust} as used in Eqs. 5.5–5.9 to account for the two horizontal forces which act at the head of the piles (pile cap level), namely the lateral pressure from the spreading soil on the pile cap (F_{cap}) and the peak inertial force (H).

Assuming a lateral pressure of 20kPa gives a lateral force per unit length acting on the pile of:

$$p_L = 20 \times D_0 = \underline{15} \text{ kN/m.} \tag{6.79}$$

The effective pile length above the fixity was calculated in Example 9 as 11.83m. The soil force acting on the pile cap in this case is given by:

$$F_{soil} = 20 \times 6 \times 1.5 = \underline{180} \text{ kN.} \tag{6.80}$$

Each of the four piles in the group is assumed to carry one quarter of the total vertical load, i.e. 2.35MN. At this axial load, the non-dimensional parameter μL is given by:

$$\mu L = \left(\sqrt{\frac{P}{EI}} \right) L = \left(\sqrt{\frac{2.35}{398}} \right) 11.83 = 0.91 \tag{6.81}$$

The factor f_Δ may then be read from Fig. 5.8 at this value of μL, giving f_Δ = 7.2. As outlined in Chapter 5, two limiting cases should be considered, namely:

1. residual response at maximum spreading displacement (post-earthquake condition); and
2. peak transient response during earthquake shaking.

The permanent horizontal displacement due to the spreading alone will first be calculated (i.e. H = 0). Using Eq. 6.78, the pile group displacement is then given by:

$$\delta_h = \frac{\left[2 \times (180 + 0) + 4 \times 15 \times 0.75 \times 11.83 \right] \times 11.83}{2 \times 4 \times 2350 \times (2 \times 7.2 - 1)} = \underline{0.042} \text{ m.} \tag{6.82}$$

During the earthquake, however, the inertial loads calculated in Example 9 will also be acting. Following the recommendations outlined in Sec. 5.2.1, the peak transient response may be estimated by superimposing the loads due to lateral spreading and inertial effects.

Equation 6.82 is therefore recalculated using H which is assumed to be the same as that computed in Example 9 (Eq. 6.76):

$$\delta_h = \frac{\left[2\times\left(180+1690\right)+4\times15\times0.75\times11.83\right]}{2\times4\times2350\times\left(2\times7.2-1\right)\times\left(11.83\right)^{-1}} = \underline{0.201}\,\text{m}. \tag{6.83}$$

Having calculated the expected peak displacements of the pile group, the pile section must be checked to ensure that the displacements are not large enough to cause yield in the piles. Comparing Eqs. 5.5 and 5.6 it is clear that for non-zero p_L, the bending moments will be larger at the bottom of the pile. The limiting displacement for pile yield may therefore be found using Eq. 5.5, by setting the moment equal to the yield moment for the pile section:

$$-1800 = -\left(2350\times\delta_{yield}\times7.2\right)-\left(\frac{15\times0.75}{0.0768^2}\times0.0695\right) \tag{6.84}$$

giving $\delta_{yield} = 0.099$m.

If the vertical load on the pile group from the superstructure was neglected, the piles would only carry at most the weight of the pile cap (assuming that there was little bearing pressure developed between the cap and the silty sand. Under these conditions:

$$P = \frac{24\times10^3\times6\times6\times1.5}{4} = \underline{324}\,\text{kN}. \tag{6.85}$$

Repeating the procedure in Eqs. 6.82 and 6.83 gives a residual displacement of $\delta_h = 0.039$m and a peak transient displacement of $\delta_h = 186$mm. For the stiff steel piles considered herein, the effects of the axial load (P-δ moments) are small. This may not be true for longer, more flexible piles, however.

Comparing the yield displacement of the pile section (Eq. 6.84) to Eqs. 6.82 and 6.83, it is clear that under the spreading forces alone, the piles would not be expected to yield, but that the high transient inertial forces are expected to cause damage in the earthquake considered in this chapter. In order to mitigate this, the peak displacements during the transient and spreading phase must be reduced. A number of steps may be taken to achieve this:

1. isolate the foundations from the inertial loads;
2. increase the number of piles in the group (increase lateral stiffness); and/or
3. increase the size of the piles (increase lateral stiffness and yield capacity).

6.5.1.1 Method 1

By substituting the yield displacement of 99mm into Eq. 6.78, it can be seen that the inertial force H must be reduced to at least 86kN (i.e. 5% of the actual inertial load). This may not be feasible in practice.

6.5.1.2 Method 2

Increasing the number of piles in the group will increase the lateral bending stiffness of the foundation, and reduce the axial load per pile (both of which will reduce displacement). However, it will also increase the area of the foundation upon which the liquefied soil pressure acts. The calculations may be repeated, increasing N in an iterative manner. Alternatively, the equations may be programmed into a spreadsheet to determine a suitable value of N. This has been completed for the foundation in this example, and the results are shown in Fig. 6.11.

It can be seen from Fig. 6.11 that a minimum of nine piles (i.e. a 3 × 3 group) would be required to prevent the piles yielding under the peak (transient inertial and kinematic) forces. This is likely to be costly (many more piles, larger pile cap).

6.5.1.3 Method 3

As an alternative to Method 2, piles of larger section may be specified. The calculations detailed previously in Example 10 have been reproduced for steel tubular piles of 1 m outside diameter (wall thickness = 16mm), which have a bending stiffness $EI = 1257MNm^2$ and plastic yield moment of 4260kNm. The results are shown in Fig. 6.12.

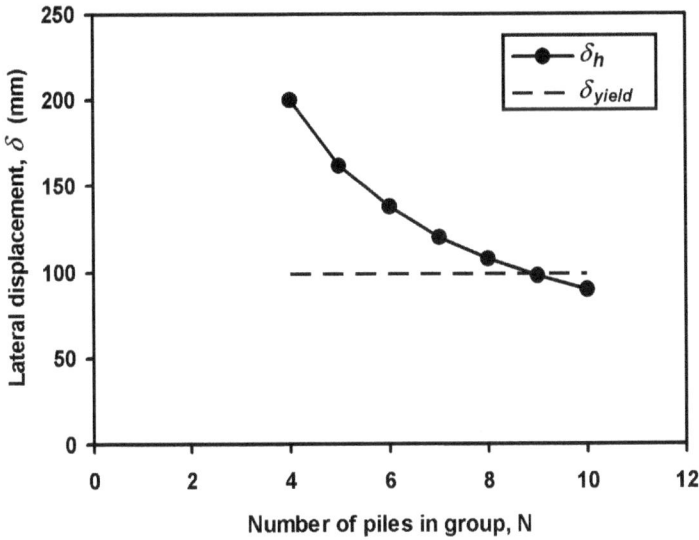

Fig. 6.11 Increase in number of piles (N) required for the pile to satisfy performance criteria (no pile yield), 0.75m diameter steel tubular piles.

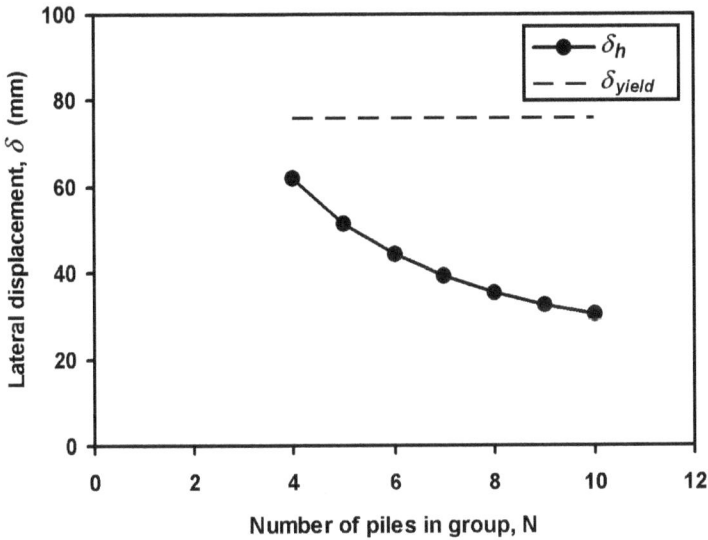

Fig. 6.12 Number of piles (N) required for the pile to satisfy performance criteria (no pile yield), 1m diameter steel tubular piles.

From Fig. 6.12, it is clear that by increasing the piles to 1m diameter sections, it is not necessary to increase the number of piles in the group. A 2×2 group of these larger piles is likely to be more economical than the 3×3 design of Method 2, requiring the pile cap to be increased only slightly in area, and using 20% less steel (in terms of total volume).

By increasing the size of the piles for the given total vertical superstructural load, the larger piles should automatically be safe against all axial conditions as the factor of safety will have increased compared to the original design consisting of the same number of smaller piles (e.g. the 2×2 group of 0.75m diameter piles in this case). The increase in bending stiffness due to the increase in diameter will also make the larger piles safer from instability (buckling) than the original design (c.f. Chapter 3).

A comparison of the different designs considered in this example is given in Table 6.5. From this table, it can be seen that the 2×2 group of 1m diameter piles outperforms the other designs, and shows how the methods outlined in this book may be used to determine an optimal, efficient pile design that is safe and/or serviceable under a given earthquake loading.

Table 6.5 Comparison of foundation designs considered in Example 10.

	Design from Example 9	Method 2 design	Method 3 design	Comments
Group configuration	2×2	3×3	2×2	
Number of piles, N	4	9	4	
Pile diameter (m)	0.75	0.75	1.00	
Pile capacity (kN)	4751	4751	6394	Using CPT/MTD
Axial load/pile (kN)	2350	1044	2350	= 9400/N
FOS (axial)	2.0	4.6	2.7	
δ_h/δ_{yield}	2.03	0.99	0.82	
Volume of steel (m^3)	2.23	5.01	3.96	
Approx size of pile cap (m)	6	9.75	8	
Comments	Unsuitable due to yield in spreading soil	Suitable, but costly	Suitable, & outperforms method 2 design in all areas	

6.5.2 Example 11: Pile group in three-layer soil profile subject to lateral spreading

In this final example, the effect of a cohesive and relatively impermeable crustal layer will be considered, which surrounds the pile cap (see Fig. 6.13). This will lie above the 8m thick medium silty sand layer and the dense clean sand layer of the previous examples. It will be assumed here that the additional overburden stress due to the crustal layer does not substantially alter the mechanical properties or liquefaction-susceptibility of the two underlying sand layers. The pile group design is the same as that considered initially in Example 10, namely a 2 × 2 group of 20m long, 0.75m diameter piles.

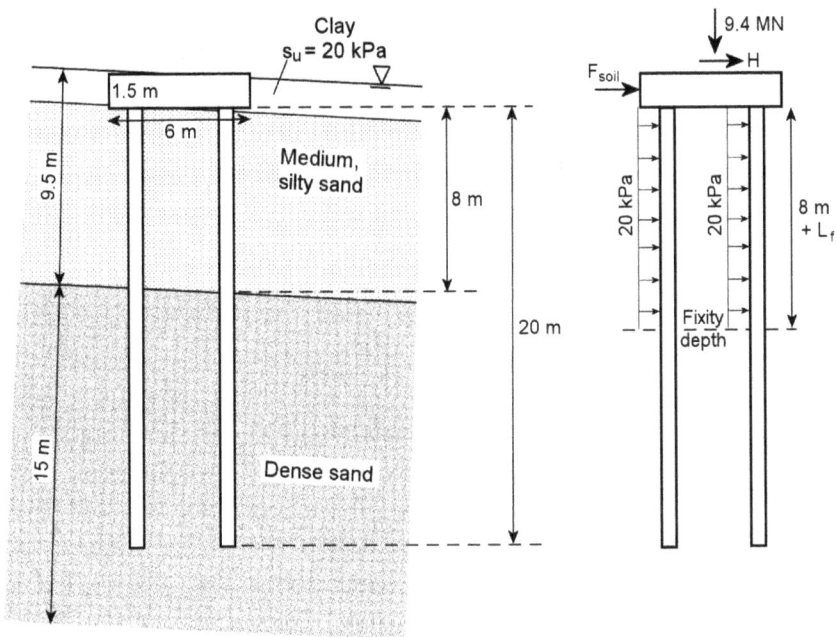

Fig. 6.13 Layout of pile group in three-layer soil profile, sloping at 3°.

The ultimate force which can now be applied by the crustal layer to the pile cap may be found using Eq. 2.17 as:

$$F_{soil} = (3\pi + 2)s_u \times 6 \times 1.5 = \underline{2056} \text{ kN.} \qquad (6.86)$$

It is assumed that the crustal layer and the liquefied sand both move further than the pile so that the limiting soil pressures in each layer act on the piles in a downslope direction. Under these loadings, the expected pile group deflections are given by Eqs. 6.87 and 6.88 for residual and peak transient conditions respectively.

$$\delta_h = \frac{\left[2 \times (2056 + 0) + 4 \times 15 \times 0.75 \times 11.83\right]}{2 \times 4 \times 2350 \times (2 \times 7.2 - 1) \times (11.83)^{-1}} = \underline{0.218} \text{ m} \qquad (6.87)$$

$$\delta_h = \frac{\left[2 \times (2056 + 1690) + 4 \times 15 \times 0.75 \times 11.83\right]}{2 \times 4 \times 2350 \times (2 \times 7.2 - 1) \times (11.83)^{-1}} = \underline{0.377} \text{ m} \qquad (6.88)$$

Comparison of the results of Eqs. 6.87 and 6.88 with Eqs. 6.82 and 6.83 confirms that the presence of nonliquefiable crustal layers which do not show appreciable strength degradation are very much more damaging to piled foundations than when they are completely surrounded by liquefied soil. Considering Eq. 6.86, this effect will become considerably worse as size of the pile cap or the shear strength of the crustal layer increases.

6.6 Summary of Inclusive Design Procedure

The examples presented in this chapter have shown how static and seismic design of pile foundations may be combined within an inclusive design method. They have shown how piles may initially be sized based on static vertical criteria, with this design then being refined and optimised to cope with seismically induced loads in nonliquefiable, liquefiable and laterally spreading soils. The process underpinning Examples 1–11 has been summarised in the form of a flow chart to aid in the design process. This is presented in Fig. 6.14. All of the evaluative procedures necessary for completion of the design may be found in Chapters 1–5 of this book (and the accompanying references), with detailed worked examples of their application to a foundation design problem presented in Examples 1 to 11.

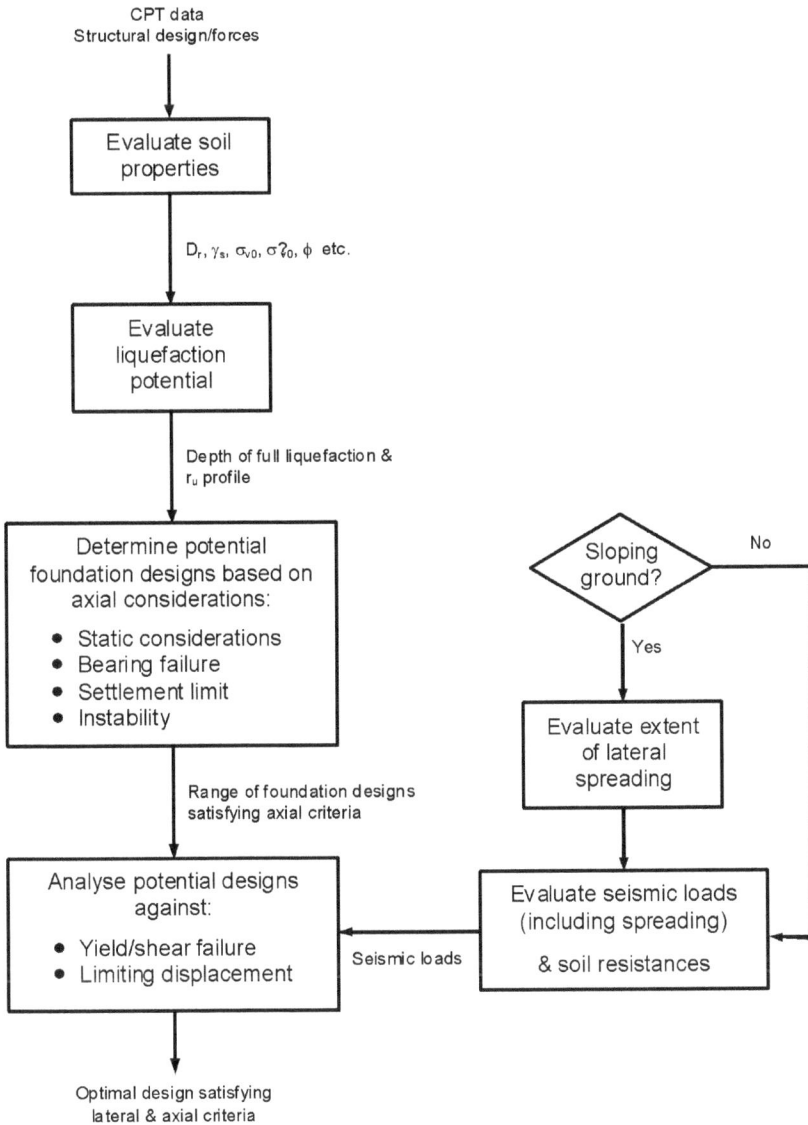

Fig. 6.14 Summary of design procedure for design of piled foundations in liquefiable soils.

References

Abdoun T., Dobry R. (2002). Evaluation of pile foundation response to lateral spreading. *Soil Dynamics and Earthquake Engineering* 22: 1051–1058.

Abdoun T. (1997). Modelling of seismically induced lateral spreading of multi-layered soil and its effect on pile foundations. PhD. thesis, Rensselaer Polytechnic Institute, Troy, New York.

Abdoun T., Dobry R., O'Rourke T.D., Goh S.H. (2003). Pile response to lateral spreads: centrifuge modelling. *J Geotech Geoenv Engng* 129(10): 869–878.

American Petroleum Institute. (1984). Recommended practice for Planning, Designing and Constructing Fixed Offshore Platforms. Code RP2A, 15th Edition, Dallas, Texas.

Architectural Institute of Japan (AIJ). (2001). *Recommendations for design of building foundations*, Japan (in Japanese)

Bartlett S.F., Youd T.L. (1995). Empirical prediction of liquefaction-induced lateral spread. *J Geotech Engng*, 121(4): 316–329.

Berezantsev V.C., Khristoforov V., Golubkov V. (1961). Load-bearing capacity and deformation of piled foundations. *Proc IV International Conference on Soil Mechanics* 2: 11–15.

Berrill J.B., Christensen S.A., Keenan R.P., Okada W., Pettinga J.R. (2001). Case Studies of Lateral Spreading Forces on a Piled Foundation. *Geotechnique* 51(6): 501–517.

Bhattacharya, S. (2003). *Pile instability during earthquake liquefaction*. PhD thesis, University of Cambridge, UK.

Bhattacharya S., Madabhushi S.P.G., Bolton M.D. (2004). An alternative mechanism of pile failure during seismic liquefaction. *Geotechnique* 54(3): 203–213.

Bhattacharya S., Bolton M.D., Madabhushi S.P.G. (2005). A reconsideration of the safety of piled bridge foundations in liquefiable soils. *Soils and Foundations* 45(4): 13–26.

Bolton M.D. (1986) The strength and dilatancy of sands. *Geotechnique* 36(1): 65–78.

Boulanger R.W., Kutter B.L., Brandenberg S.J. Singh P., Chang D. (2003). *Pile foundations in liquefied and laterally spreading ground during earthquakes: centrifuge experiments & analyses*. Centre for Geotechnical Modelling Report No. UCD/CGM-03/01, University of California at Davis, USA.

Brandenberg S.J., Boulanger R.W., Kutter B.L. (2005). Discussion of – Single piles in lateral spreads: field bending moment evaluation. *J Geotech Geoenv Engng* 131(4): 529–531.

Brandenberg S.J., Boulanger R.W., Kutter B.L., Chang D. (2005). Behaviour of pile foundations in laterally spreading ground during centrifuge tests. *J Geotech Geoenv Engng* 131(11): 1378–1391.

Broms B. (1966). Methods of calculating the ultimate bearing capacity of piles, A summary. *Sols-Soils* 5(18–19): 21–31.

Budhu M., Davies T.G. (1987). Nonlinear analysis of laterally loaded piles in cohesionless soils. *Canadian Geotechnical J* 24: 289–296.

Budhu M., Davies T.G. (1988). Analysis of laterally loaded piles in soft clays. *J Geotech Engng* 114(1): 21–39.

Castro G. (1975). Liquefaction and cyclic mobility of saturated sands. *J Geotech Engng* 101(6): 551–569.

Castro G., Poulos S.J. (1977). Factors affecting liquefaction and cyclic mobility. *J Geotech Engng* 103(6): 501–516.

Chang D., Boulanger R.W., Kutter B.L., Brandenberg S.J. (2005). Inertial and spreading load combinations of soil-pile-structure system during liquefaction-induced lateral spreading in centrifuge tests. *Proc 16th Int Conf on Soil Mechanics and Geotechnical Engineering* 2: 1963–1966.

Cheng Y.M. (2004). N_q factor for pile foundations by Berezantzev. *Géotechnique* 54(2): 149–150.

Chow F.C. (1997). Investigations in the behaviour of displacement piles for offshore foundations. PhD thesis, Imperial College, London.

Coelho P.A.L.F., Haigh S.K., Madabhushi S.P.G., O'Brien A. (2003). Centrifuge modelling of earthquake-induced liquefaction in loose uniform deposits. *Proc BGA International Conference on Foundations: Innovations, Observations, Design and Practice.*

Coelho P.A.L.F., Haigh S. K., Madabhushi S.P.G. (2003). Boundary effects in dynamic centrifuge modelling of liquefaction in sand deposits. *Proc. 16th ASCE Engineering Mechanics Conference.*

Coelho P.A.L.F., Haigh S.K., Madabhushi S.P.G., O'Brien A.S. (2007). Post-earthquake behaviour of footings when using densification as a liquefaction resistance measure. *Ground Improvement* 11(1): 45–53.

Coyle H.M., Castello R.R. (1981). New design correlations for piles in sand. *J Geotech Engng* 197(7): 965–985.

Davies T.G., Budhu M. (1986). Nonlinear analysis of laterally loaded piles in heavily overconsolidated clays. *Geotechnique* 36(4): 527–538.

De Alba P.A. (1983). Pile settlement in liquefying sand deposit. *J Geotech Engng* 109(9): 1165–1179.

Dobry R., Abdoun T. (1998). Post-triggering response of liquefied sand in the free field and near foundations. *Proc 3rd ASCE Specialty Conference on Geotechnical Engineering and Soil Dynamics.* 270–300.

Dobry R., Abdoun T., O'Rourke T.D., Goh S.H. (2003). Single piles in lateral spreads: field bending moment evaluation. *J Geotech Engng* 129(10): 879–889.

Dobry R., Vucetic M. (1991). Effect of Soil plasticity on cyclic response. *J Geotech Engng ASCE* 117(1): 89–107.

Doroudian M., Vucetic M. (1999). Results of Geotechnical laboratory tests on soil samples from the UC Riverside Campus, UCLA Research Report No. ENG-99-203. University of California, Los Angeles.

Elgamal A., Yang Z., Lai T., Kutter B.L., Wilson D.W. (2005). Dynamic response of saturated dense sand in laminated centrifuge container. *J Geotech. Engng* 131(5):598–609.

Eurocode 7 Geotechnical Design – *Part 1 (1997) General rules*, CEN European Committee for Standardisation, BS EN 1997-1:2004.

Eurocode 8 – Part 1 (2003) Design provisions for earthquake resistance of structures – General rules seismic actions and rules for buildings, CEN European Committee for Standardisation, prEN 1998-5:2003.

Eurocode 8 – Part 5 (2003) Design provisions for earthquake resistance of structures – foundations, retaining structures and geotechnical aspects. CEN European Committee for Standardisation, prEN 1998-5:2003.

Fellenius B.H. (1972). Downdrag on long piles in clay due to negative skin friction. *Canadian Geotech. J* 9(4): 323–337.

Fleming W.G.K. (1992). A new method for single pile settlement prediction and analysis. *Geotechnique* 42(3): 411–425.

Florin V.A., Ivanov P.L. (1961). Liquefaction of saturated sandy soils. *Proc 5th International Conference on Soil Mechanics and Foundation Engineering* 1: 107–111.

Fukuoka M. (1966). Damage to Civil Engineering Structures. *Soils and Foundations*, 6(2): 45–52.

Gabr M.A., Wang J.J., Zhao M. (1997). Buckling of piles with general power distribution of lateral subgrade reaction. *J Geotech. Engng* 123(2): 123–130.

Gazetas G. (1984). Seismic response of end-bearing piles. *J Soil Dynamics and Earthquake Eng* 3(2): 82–93.

Gazetas G. (1990). Foundation vibrations in H-Y Fang (ed), *Foundation Engineering Handbook, 2nd Edition*, pp. 553–593. Van Nostrand Reinhold Co.

Gazetas G. (1991). Formulas and charts for impedances of surface and embedded foundations. *J Geotech. Engng* 117(9): 1363–1381.

Gazetas G., Fan K., Kaynia A., Kausel E. (1991). Dynamic interaction factors for floating pile groups. *J Geotech. Engng* 117: 1531–1548.

Ghosh B., Madabhushi S.P.G. (2002). An efficient tool for measuring shear wave velocity in the centrifuge. *Proc International Conference on Physical Modelling in Geotechnics*.

Goh S., O'Rourke T.D. (1999). Limit state model for soil-pile interaction during lateral spread. *Proc 7th U.S.–Japan Workshop on Earthquake Resistant Design of Lifeline Facilities and Countermeasures Against Soil Liquefaction*.

Haigh S.K., Madabhushi S.P.G. (2005). The Effects of Pile Flexibility on Pile Loading in Laterally Spreading Slopes in R W Boulanger & K Tokimatsu (eds), *Seismic Performance and Simulation of Pile Foundations in Liquefied and Laterally Spreading Ground,* pp. 24–37. ASCE Geotechnical Special Publication No. 145.

Haigh S.K. (2002). *Effects of Liquefaction on Pile Foundations in Sloping Ground.* PhD thesis, University of Cambridge, UK.

Haigh S.K., Madabhushi S.P.G., Soga K., Taji Y., Shamoto Y. (2000). Lateral spreading during centrifuge model earthquakes. *Proc Geo Eng 2000.*

Haigh S.K., Madabhushi S.P.G., Soga K., Taji Y., Shamoto Y. (2001). Analysis of liquefied flow in centrifuge model earthquakes. *Proc 4th Int Conf on Recent Advances in Soil Dynamics and Geotechnical Earthquake Engineering.*

Hall J.F. (1984). Forced vibration and earthquake behaviour of an actual pile foundation. *J Soil Dynamics and Earthquake Eng* 3(2): 94–101.

Hamada M., Yasuda S., Isoyama R., Emoto K. (1986). *Study on liquefaction induced permanent ground displacements.* Association for the Development of Earthquake Prediction in Japan.

Hamada M. (1992). Large ground deformations and their effects on lifelines in M. Hamada and T.D. O'Rourke (eds), *Case studies of liquefaction and lifeline performance during past earthquakes,* Japanese Case Studies, Technical Report NCEER-92-0001, Vol.1.

Hardin B.O., Drnevich V.P. (1972). Shear modulus and damping in soils: design equations and curves. *J Geotech. Engng* 98(7): 667–692.

Horne J.C., Kramer S.L. (1998). *Effects of liquefaction on pile foundations.* US Dept of Transportation, Federal Highway Administration Report No. WA-RD 430.1

Idriss I.M., Boulanger, R.W. (2004). Semi-empirical procedures for evaluating liquefaction potential during earthquakes. *Proc 11th Internatonal Conference on Soil Dynamics and Earthquake Engineering* 1: 32–56.

Ishihara K. (1993). Liquefaction and flow failure during earthquakes. *Géotechnique* 43(3): 351–415.

Ishihara K. (1997). Terzaghi oration: Geotechnical aspects of the 1995 Kobe earthquake *Proc ICSMFE,* pp. 2047–2073.

Iwasaki T., Tatsuoka F., Tokida K., Yasuda S. (1978). A Practical Method for Assessing Soil Liquefaction Potential based on Case Studies at Various Sites in Japan. *Proc 2nd International Conference on Microzonation, San Francisco* 2: 885 –896.

Japan Road Association (JRA). (2002). Specifications for highway Bridges, part V: seismic design, Japan (in Japanese).

Jardine R.J., Chow F.C. (1996). *New design methods for offshore piles,* MTD Publications, No. 96/103, London: Marine Technology Directorate.

Knappett J.A. (2006). Piled foundations in liquefiable soils: accounting for axial loads. PhD thesis, University of Cambridge, UK.

Knappett J.A. (2008). Discussion of – Design charts for seismic analysis of single piles in clay. *Geotechnical Engineering* 161(2): 115–116.

Knappett J.A., Madabhushi S.P.G. (2008). Liquefaction-induced settlement of pile groups in liquefiable and laterally-spreading soils. *J Geotech. Engng* 134(11):

Knappett J.A., Madabhushi, S.P.G. (2008). Designing against pile tip bearing capacity failure in liquefiable soils. *Proc 2nd BGA Int. Conf on Foundations* 2: 1237–1246.

Knappett J.A., Madabhushi S.P.G. (2005). Modelling of Liquefaction-Induced Instability in Pile Groups in R.W. Boulanger & K. Tokimatsu (eds), *Seismic Performance and Simulation of Pile Foundations in Liquefied and Laterally Spreading Ground*, pp. 255–267. ASCE Geotechnical Special Publication No. 145.

Kokusho T. (1999). Formation of water film in liquefied sand and its effect on lateral spread. *J Geotech. Geoenv Engng* 125(10): 817–826.

Kong F.K., Evans R.H. (1987). *Reinforced and prestressed concrete* E&FN Spon, London.

Kutter B.L., Balakrishnan A. (1999). Visualization of soil behavior from dynamic centrifuge model tests. *Proc 2nd Int. Conf on Earthquake Geotechnical Engineering* 3: 857–862.

Liu L., Dobry R. (1995). Effect of liquefaction on lateral response of piles by centrifuge model tests NCEER report submitted to FHWA.

Madabhushi S.P.G. (2004). Geotechnical Aspects of the 921 Ji-Ji earthquake of Taiwan, in *EEFIT Report on 921 Ji-Ji Earthquake*, Institution of Structural Engineers, London, UK.

Madabhushi S.P.G., Patel D., Haigh S.K. (2005). Geotechnical Aspects of the Bhuj Earthquake, in *EEFIT Report on the Bhuj Earthquake*, Institution of Structural Engineers, London, UK.

Makris N., Gazetas G. (1992). Dynamic pile-soil-pile interaction. Part II: Lateral and seismic response. *J Earthquake Eng and Structural Dynamics* 21: 145–162.

Makris N., Gazetas G., Delis E. (1996). Dynamic soil-pile-foundation-structure interaction: records and predictions. *Geotechnique* 46(1): 33–50.

Malvick E.J., Kutter B.L., Boulanger R.W., Kulasingham R. (2006). Shear localization due to liquefaction-induced void redistribution in a layered infinite slope. *J Geotech. Engng* 132(10): 1293–1303.

Matesic L., Vucetic M. (1998). Results of Geotechnical laboratory tests on soil samples from the UC Riverside Campus, UCLA Research Report No. ENG-98-198 University of California, Los Angeles.

Matlock H., Reese L.C. (1960). Generalised solutions for laterally loaded piles. *J Geotech. Engng* 86(5): 63–91.

McClelland B., Focht J.A. (1958). Soil modulus for lateral loaded piles. *Trans ASCE*, 123: 1046–1063.

Meyerhof G.G., Valsangkar A.J. (1977). Bearing capacity of piles in layered soils. *Proc VIII Intl Conf Soil Mech. Foundation Eng* 1: 645–650.

Meyersohn W.D. (1994). *Pile response to liquefaction-induced lateral spread.* PhD thesis, Cornell University, NY.

Motoki K., Noriaki S., Ryosuke U. (2008). Progressive damage simulation of foundation pile of the Showa bridge caused by lateral spreading during the 1964 Niigata earthquake, *Proc 2nd Intl Conf on Geotechnical Engineering for Disaster Mitigation and Rehabilitation.*

Murff J.D., Hamilton J.M. (1993). P-Ultimate for undrained analysis of laterally loaded piles. *J Geotech. Engng* 119(1): 91–107.

Nemat-Nasser S., Shokooh A. (1979). A unified approach to densification and liquefaction of cohesionless sand in cyclic shearing. *Canadian Geotechnical Journal* 16(4): 659–678.

Newmark N.M. (1965). Effects of earthquakes on dams and embankments. *Géotechnique* 15(3): 139–160.

Novak M. (1991). Piles under dynamic loads: State of the Art. *Proc 2nd Int. Conf on Recent Advances in Geotechnical Earthquake Engineering and Soil Dynamics* 3: 250–273.

Pamuk A., Zimmie T.F., Abdoun T. (2003). Pile group foundations subjected to seismically induced lateral spreading. *Proc. 1st BGA Int. Conf. on Foundations* 715–722.

Pender M.J. (1993). A seismic pile foundation design analysis. *Bulletin of New Zealand National Soc. for Earthquake Engineering* 26(1): 49–174.

Pender M.J. (1996). Earthquake resistant design of foundations. *Bulletin of New Zealand National Soc. for Earthquake Engineering* 29(3): 155–171.

Permanent International Association for Navigation Association (PIANC). (2001). *Seismic design guidelines for port structures* Balkema, Rotterdam.

Poulos H.G., Davis E.H. (1980). *Pile Foundation Analysis and Design* John Wiley and Sons.

Pyke R. (1979). Nonlinear soil models for irregular cyclic loadings. *J Geotech. Engng* 105(6): 715–726.

Railway Technical Research Institute (RTRI). (1999). *Design standard for railway facilities – seismic design* (in Japanese).

Ramos R., Abdoun T., Dobry R. (1999). Centrifuge Modelling of Effect of Superstructure Stiffness on Pile Bending Moments due to Lateral Spreading. *Proc 7th US–Japan Workshop on Earthquake Resistant Design of Lifeline Facilities and Countermeasures Against Soil Liquefaction* 599–608.

Randolph M.F., Houlsby G.T. (1984). The limiting pressures on a circular pile loaded laterally in cohesive soil. *Geotechnique* 34(4): 613–623.

Randolph M.F. (2003). Science and empiricism in pile foundation design. *Geotechnique* 53(10): 847–875.

Randolph M.F., Dolwin J., Beck R.D. (1994). Design of driven piles in sand. *Geotechnique* 44(3): 427–448.

Rauch A.F., Martin J.R. (2000). EPOLLS model for predicting average displacements on lateral spreads. *J Geotech. Engng* 126(4): 360–371.

Riks E. (1972). The application of Newton's method to the problem of elastic stability. *J Appl Mech.* 39: 1060–1065.

Riks E. (1979). An incremental approach to the solution of snapping and buckling problems. *Int J Solids & Struct* 15: 529–551.

Rollins K.M., Strand S.R. (2006). Downdrag forces due to liquefaction surrounding a pile. *Proc 8ᵗʰ National Conference on Earthquake Engineering.*

Seed H.B., Idriss I.M. (1971). Simplified procedure for evaluating soil liquefaction potential. *J Geotech. Engng* 97(9): 1249–1273.

Seed H.B., Martin P.P., Lysmer J. (1975). The generation and dissipation of pore water pressures during soil liquefaction. *Report No. UCB/EERC-75/26*, University of California at Berkeley, USA.

Sitharam T.G., Govindaraju L. (2004). Geotechnical aspects and ground response studies in Bhuj earthquake, India. *Geotechnical and Geological Engineering* 22: 439–455.

Soga K. (1997). Geotechnical aspects of Kobe earthquake in *EEFIT report on the Kobe Earthquake*, Institution of Structural Engineers, UK.

Suzuki H., Tokimatsu K., Sato M., Abe A. (2005). Factors affecting Horizontal Subgrade Reaction of Piles During Soil Liquefaction and Lateral Spreading, in R.W. Boulanger & K. Tokimatsu (eds), *Seismic Performance and Simulation of Pile Foundations in Liquefied and Laterally Spreading Ground.* ASCE Geotechnical Special Publication No. 145, pp. 1–10.

Takahashi A., Kuwano J., Arai Y., Yano A. (2002). Lateral resistance of buried cylinder in liquefied sand. *Proc Int Conf on Physical Modelling in Geotechnics: ICPMG '02*, 477–482.

Takata, T., Tada, Y., Toshida, I., Kuribayashi, E. (1965). Damage to bridges in Niigata earthquake. Report No. 125-5, Public Works Research Institute, Japan (in Japanese).

Tamura S., Tokimatsu K. (2005). Seismic earth pressure acting on embedded footing based on large-scale shaking table tests in R.W. Boulanger & K. Tokimatsu (eds), *Seismic Performance and Simulation of Pile Foundations in Liquefied and Laterally Spreading Ground.* ASCE Geotechnical Special Publication No. 145, pp. 83–96.

Timoshenko S.P., Gere J.M. (1961). *Theory of elastic stability* McGraw-Hill Book Company Inc., New York.

Tokimatsu K, Mizuno H., Kakurai M. (1996). Building Damage associated with Geotechnical problems. *Soils and Foundations* Special Issue:219–234.

Tokimatsu K., Asaka Y. (1998). Effects of liquefaction-induced ground displacements on pile performance in the 1995 Hyogoken-Nambu Earthquake. *Soils and Foundations* Special Issue: 163–177.

Tokimatsu K., Suzuki H., Sato M. (2004). Influence of inertial and kinematic components on pile response during earthquakes. *Proc. 11th Int. Conf. on Soil Dynamics and Earthquake Engineering* 1: 768–775.

Tomilinson M.J. (1986). *Foundation Design and Construction*, Longman Scientific & Technical, England.

Towhata I., Orense R.P., Toyota H. (1999). Mathematical principles in prediction of lateral ground displacement induced by seismic liquefaction. *Soils and Foundations* 39(2): 1–19.

Towhata I., Vargas-Monge W., Orense R.P., Yao M. (1999). Shaking table tests on subgrade reaction of pipe embedded in sandy liquefied subsoil. *Soil Dynamics & Earthquake Engng* 18: 347–361.

Uchida A., Tokimatsu K. (2005). Comparison of current Japanese Design Specifications for Pile Foundations in Liquefiable and Laterally Spreading Ground in R W Boulanger & K Tokimatsu (eds), *Seismic Performance and Simulation of Pile Foundations in Liquefied and Laterally Spreading Ground*. ASCE Geotechnical Special Publication No. 145, pp. 61–70.

Vardanega P.J., Bolton M.D. (2009). Stiffness of clays and silts, submitted to *Geotechnique*.

Vesic A.S. (1972). Expansion of cavities in infinite soil mass. *J Geotech Engng* 98(3): 265–289.

Vesic A.S. (1967). *A study of bearing capacity of deep foundations*, Final Report, Proj. B-189, Atlanta, Georgia Institute of Technology.

Vesic A.S. (1970). Tests on instrumented piles, Ogeechee river site, *J Geotech Engng* 96(2): 561–584.

Vijayvergiya V.N. (1977). Load-movement characteristics of piles. *Proc. Ports77*.

Wolf J.P. (1985). *Dynamic soil-structure interaction*, John Wiley and Sons, New York.

Yang Z., Elgamal A., Parra E. (2003). Computational model for cyclic mobility and associated shear deformation. *J Geotech Engng* 129(12): 1119–1127.

Yoshida N., Tazoh T., Wakamatsu K., Yasuda S., Towhata I., Nakazawa H., Kiku H. (2007). Causes of Showa bridge collapse in the 1964 Niigata earthquake based on eye-witness testimony. *Soils and Foundations* 47(6): 1075–1087.

Youd T.L., Hansen C.M., Bartlett S.F. (2002). Revised multilinear regression equations for prediction of lateral spread displacement. *J Geotech Engng* 128(12): 1007–1017.

Zhao Y. (2008). *In Situ Soil Testing for Foundation Performance Prediction*. PhD thesis, University of Cambridge, UK.

Index

www.ingramcontent.com/pod-product-compliance
Lightning Source LLC
Chambersburg PA
CBHW050558190326
41458CB00007B/2091

*9 7 8 1 8 4 8 1 6 3 6 2 1 *